Digital
Home
Recording

Editor, Carolyn Keating;
and Technical Editor,
Craig Anderton

MF Miller Freeman Books

Published by Miller Freeman Books
600 Harrison Street, San Francisco, CA 94107
Publishers of *Keyboard, Bass Player,* and *Guitar Player* magazines

un Miller Freeman
A United News & Media publication

Distributed to the book trade in the U.S. and Canada by
Publishers Group West, P.O. Box 8843, Emeryville, CA 94662

Distributed to the music trade in the U.S. and Canada by
Hal Leonard Publishing, P.O. Box 13819, Milwaukee, WI 53213

Cover Design: Audrey Welch
Design and Typesetting: Greene Design

Library of Congress Cataloging in Publication Data:

Digital home recording : editor, Carolyn Keating, technical editor, Craig Anderton.
     p. cm.
  ISBN 0-87930-380-8
  1. Sound—Recording and reproducing—Digital techniques. 2. Sound
  studios. I. Keyboard
TK7881.65.D54 1998
621.389'32—DC21   98-41676
                                                                CIP

Printed in the United States of America
98 99 00 01 02 03   5 4 3 2 1

CONTRIBUTORS

Compiled from articles written by Jim Aikin, Craig Anderton, Michael Babcock, Eddie
Ciletti, Julian Colbeck, Scott Garrigus, Chris Gill, Joe Gore, Ted Greenwald, David
Miles Huber, Brent Hurtig,Mike Hurwicz, Jeff Klopmeyer, Robert Lauriston, Michael
Marans, Howard Massey, Roger Nichols, Dave O'Neal, Martin Polon, Greg Rule,
Marvin Sanders, Bennet Spielvogel, Mark Vail, Guy Wright, and originally published
in *Keyboard, Music & Computers,* and *EQ.*

CONTENTS

INTRODUCTION

As little as five years ago, if you wanted to make an album, you had to go into a recording studio, pay an engineer and/or producer a lot of money, and come away with a heavy and unplayable-in-your-home two-inch tape and a DAT. Could your friends, family, and fans hear this album you spent so much time and money recording? No—not in those formats. Most people don't have 24-track tape machines in their living rooms, and since DAT died as a commercial medium, they probably don't have a DAT machine either. But you could give them a cassette, and most likely did. And under all the hiss and tape wow and flutter, they would hear your magnum opus. But not in its best possible manifestation—clean, clear, noiseless digital. In order to do that you gave the DAT to a mastering facility so they could press it into a CD— after you dished out another chunk of change. *Then* you could give the CD to your loved ones. Well, all that has changed in recent years. Prices on hard disk drives and computer memory have dropped drastically, computers are more affordable than ever, and digital recording technology in general is now well within the price range of the home studio musician. But the most exciting thing is that now you can burn your CDs at home! CD-R drives have plummeted from as much as $100,000 (1989) to as little as $400 today. Now you can be the engineer, producer, and manufacturer of your CDs, and you can take all the time you need without watching your money (aside from your initial equipment investments) fly out of your pocket. And your final product will have the best possible fidelity. It's a no-brainer. Naturally, you've decided you want to make home recordings using this newly affordable digital media. Well, this book is geared toward you—the home digital recordist, who is perhaps a little confused about what exactly they are going to need to put their digital studio together, isn't quite certain what their options are, and is a little rusty on some of the finer points of recording in general and digital audio in particular. Relax. This book is intended to answer your questions, point you toward the appropriate gear, and hopefully expand your knowledge of digital audio. And from this you should be able to put your studio together and make that phenomenal CD. Compiled from articles published in *Keyboard*, *Music & Computers*, *EQ*, *Guitar Player*, and *Interactivity* magazines, the chapters within were written by some of the most knowledgeable people in the business. It is intended to be a comprehensive yet understandable guide to digital recording, covering everything from the nature of digital audio to studio setups, equipment descriptions, and the process of recording and burning a CD.

Speaking of which, I'm sure you must be anxious to do just that. So let's get started. The fun is about to begin.

—*Carolyn Keating*

ACKNOWLEDGMENTS

A book of this kind would not be possible without the contributors, all of whom I'd like to thank for their ability to write about a complex, technical subject in a comprehensive and understandable way (with even humor, at times!). It is my hope that I have integrated and edited their writings in an equally clear fashion.

Particular thanks goes to Matthew Kelsey at Miller Freeman, who had the inspired idea for this book, and the faith and flexibility necessary to allow me to put it together. I am also indebted to Dorothy Cox at Miller Freeman, who was instrumental in setting me up for this project. Further gratitude goes to David Earl at Leo's Pro Audio, who was gracious enough to take the time out from his busy schedule to answer my questions.

Additional thanks to Jan Hughes, Jeff Campbell, Joanna Hurley, and especially my partner Lisa, who put up with the magazines strewn about the floor, calmed my deadline anxieties, and enabled me to see the light at the end of the tunnel.   —CK

ABOUT THE EDITORS

Carolyn Keating is a freelance editor, audio engineer, and musician. She earned her audio engineering certificate from the California Recording Institute in Menlo Park. Along with stints at record labels and production companies, she has worked for *Bay Area Musician (BAM)* and *MicroTimes*, and edited numerous music publications. She is currently in the process of finishing a CD.

Technical Editor Craig Anderton, who coined the term "electronic musician," is a musician and author of various books, including *Home Recording for Musicians*, *Do-It-Yourself Projects for Guitarists*, *Electronic Projects for Musicians*, and *MIDI for Musicians*. Craig has also written numerous articles for magazines such as *Guitar Player*, *Keyboard*, *EQ*, *Rolling Stone*, *Mix*, and *Byte*. In addition to serving as consulting editor to *Guitar Player* and technology editor to *EQ*, Craig lectures all over the world, consults to manufacturers in the music business, and is responsible for some of the sounds you hear coming out of various instruments, as well as some of their design features.

The World of Digital Audio

Your mother thinks your home studio looks like the bridge of the starship *Enterprise.*You try to tell her, "No, Mom, it's really simple. See, the computer handles the MIDI sequencing and tapeless sound recording. That's a 1.5-gigabyte hard drive for the digital audio, and those points in the patch bay are the effects sends from the mixer, and this is how you enter edit mode on the JD-990, and. . . ." Her eyes wander. She mentions that you forgot to water the fern.

You should know better. It *is* simple—once you've spent years getting familiar with it all. But do you know how the technology that you use every day works? Or do you just scrunch your eyes shut, hit the power switch, and hope nothing blows up?

There's nothing wrong with just hooking up a few gadgets and concentrating on the music. That may even be more productive than getting lost in the digital ozone. However, there are important advantages to knowing a little about what's going on behind the front panel. Like getting the most out of your equipment. Like not wasting money buying things that don't meet your needs. And, possibly, impressing your friends by dropping technical jargon into casual conversations.

Most of us, even if we've been wallowing in electronic music technology for years, have a few blank cards in our fact files. If you'd like to test your expertise and maybe stretch it a little, read on. If you're new (or comparatively new) to the whole process of making music with digital hardware and software, we'll try to get you up to speed in some basic areas that may come in handy down the line. And if you're a whiz at computers but weak in the MIDI department, or vice-versa, this is your chance to become better rounded.

The Fundamentals of Sound

Getting the most out of any electronic music system requires a basic understanding of acoustics—the physics of sound. (For more in-depth coverage, investigate further at your local library or check the bibliography at the back of the book.)

Sound consists of rapid variations in air pressure. A vibrating string creates these changes directly: Twang it, and it moves back and forth, pushing the air molecules first one way, then another. A speaker does much the same thing when driven by an electrical current: The speaker cone moves in and out, displacing a certain amount of air.

These waves of air pressure, or *sound waves*, travel through the air at about 1,000 feet per second (the speed of sound). In the conventional diagram of a sound wave (see Figure 1), high points indicate zones of greater pressure, while low points show zones of lesser pressure.

❶ *Sound is composed of waves of compression and expansion. The conventional diagram of a sound wave represents zones of higher pressure (compression) as hills and zones of lower pressure (expansion) as valleys.*

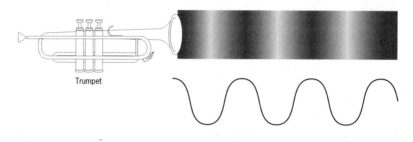

Trumpet

Our ears are most sensitive to sound waves that vibrate at more than 20 cycles per second and less than 20,000 cycles per second. A cycle typically includes both the positive and negative part of a wave; "cycles per second" is abbreviated as Hertz (Hz), and measures the sound's frequency. A thousand cycles is called a *kilohertz*, or kHz for short. The range from 20Hz to 20kHz is commonly considered the range of human hearing, though older people, or people of any age who abuse their hearing, often suffer hearing loss in the upper range—say, above 8–10kHz.

Overtones. A plucked string doesn't simply vibrate as a unit, producing sound waves of one pure frequency. Instead, it vibrates in a complex way that includes vibrations that are mathematical multiples of the basic frequency. Essentially, the whole string is vibrating, the halves of the string are vibrating, the thirds of the string are vibrating, and so on. All of these vibrations occur in the same string at the same time; the tone that we perceive as coming from the plucked string includes vibrations of all these different frequencies.

The frequency of the whole string vibrating is called the *fundamental*. The higher frequencies at which the same string also vibrates are called its *overtones*. The fundamental and its overtones are related mathematically: If the fundamental is at 100Hz, for example, the overtones may fall at 200Hz, 300Hz, 400Hz, 500Hz, and so on (the exact frequencies depend on the string's overtone structure). Usually the higher overtones will be lower in volume than the lower overtones. You'll also hear the terms *harmonics* and *partials* used. Technically, the fundamental is the first partial, so the second partial is the first overtone. Any component of a sound, whether or not it has the mathematical relationship we've just described, is a partial. With sounds that exhibit this type of mathematical coherence, a harmonic is the same thing as a partial.

In different sound sources (violin, pipe organ, the human voice), the fundamental and various overtones are mixed in various proportions—some louder and some softer. It's the mix of overtones that gives each instrument sound its characteristic *timbre* (pronounced "TAM-br"), or sound quality.

Some acoustic instrument partials are "out of tune" with respect to the fundamental rather than vibrating at the frequencies predicted by mathematical theory. An extreme example is a church bell, whose fundamental and first overtone are usually separated by an interval of about a major sixth (from 100Hz to about 166Hz, for example). Partials that are not whole-number multiples of the fundamental are called *clangorous*, because they resemble the sound of a bell.

Fourier Analysis. A Frenchman named Fourier (pronounced "foor-yay") proposed a method of mathematical analysis to describe any sound, clangorous or not, in terms of its fundamental and overtones. Fourier analysis considers every sound as containing some number of *sine waves* that are vibrating at different frequencies. (A sine wave is a pure tone with no overtones.) Because few sounds in nature are static, in Fourier analysis each sine wave also changes in amplitude (loudness) over time.

The concept that a sound consists of many components at different frequencies is important to many processes involving electronic music. For instance, when using an equalizer (see "Effects Processors"), you boost or cut the amplitude of some frequency band—for example, 4kHz. This has an audible effect only if the original sound feeding the equalizer contains overtones around 4kHz. If most of its overtone energy is at 1kHz, boosting at 4kHz won't do much. Another example: Some synthesizer patches use a second oscillator tuned several octaves above the first (and usually set to a lower output level) to add a specific overtone to the timbre.

A synthesizer *filter* shapes a sound's overtone content (usually this is under the control of an *envelope generator*, which changes the overtone content over time). When a lowpass filter removes the higher overtones from the latter part of each note, it mimics the response of an acoustic instrument (especially a struck or plucked instrument like piano or guitar) in which physical sound energy dissipates gradually, starting with the upper overtones.

Analog Audio

Compared to the world of digital audio, the old-fashioned analog kind seems intuitively obvious: Just plug in your instrument, crank the volume knob, and go. Nonetheless, how you hook up various devices can definitely influence your music's sound quality.

Impedance. Signal level is one of the most important issues we need to cover, but to understand it properly, we need to cover some basic audio principles such as impedance and decibels.

Ohms measure *impedance*, which generally stands for opposition, or resistance, to current flow in an electrical circuit. Most audio equipment has a very low output impedance so that the signal goes through the minimum possible resistance, thus producing the strongest possible signal. The input impedance, on the other hand, measures the resistance an incoming signal sees to ground. Here you want a relatively high impedance, so that as little of the signal as possible goes to ground, and can proceed directly to the next input stage. Figure 2 summarizes how input and output impedance affect the signal. Incidentally, gear designed for guitar, bass, and other instruments using passive pickups will have an extremely high input impedance to avoid loading down the pickups, whose output impedance is higher than most other gear.

High-impedance equipment usually has a ¼" phone plug. Feeding a high impedance output, such as a guitar, into a low impedance input, such as a mic input on a mixer, requires a transformer called a *direct box*. This lowers the signal to mic level

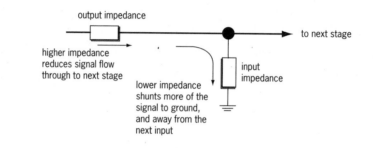

❷ *How imped-ance affects signal strength as it flows from one stage to the next.*

and matches impedances. (Note that there are other types of direct boxes, such as devices that provide gain, add tone controls, use active circuitry instead of a transformer, etc.)

Decibels. A *decibel* (abbreviated dB) is one tenth of a unit called a *Bel* (named after Alexander Graham Bell), which was originally developed to measure the power of various signals in telephone systems. A decibel is not an absolute measurement like pounds or centimeters, but is a *ratio* between two numbers, one of which is a standard reference value. Without knowing what the reference is, the number is meaningless.

Most people think of decibels as a measure of volume, where 0dB is referenced as the threshold of hearing, and 130dB is called the threshold of pain because the sound is so loud it hurts. But there are different types of decibels, as indicated by a single-letter suffix (e.g., dBv, dBm, dBu, etc.), which reference the signal being measured to different standardized levels. For example, the dBm, which describes a signal's power or wattage, is referenced to 1 milliwatt of power. In audio equipment, 1 milliwatt equals 0.775 volts into a 600-ohm input impedance, and with the dBm scale, equals 0dBm.

Since not all audio gear is based on changes in power (or 600-ohm output and input impedances), another scale was invented based on *voltage*, the dBV. This is referenced to 1 volt, so that 0dBV = 1 volt. The dBu scale (used to eliminate the 2.2dB difference between the dBV and dBm scales) determines voltage independent of impedance. It is simply referenced to 0.775 volts (so 0dBu = 0.775 volts, regardless of impedance).

Signal Levels. There are also more qualitative ways to define levels. Microphones put out a *mic-level* signal (around −60dBV). This low-level signal requires a preamp to boost it to *line level.* Passive guitar pickups generate signals that are stronger than mics, but weaker than electronic devices such as synthesizers. Typically, synthesizer outputs and mixer inputs are line-level; this level is nominally −10dBV, though in fact synth outputs tend to hover around 0dBu. Professional devices quite often work with higher-level, +4dBm signals.

One ramification of using devices based on different nominal levels is that if you feed a +4dBm signal into an input designed for −10dBV signals, the +4dBm signal will be too strong, and will cause distortion unless the −10 dBV input includes a *pad*, a switch or knob that lowers the incoming signal's level before it reaches the rest of the circuitry. Going from a −10dBV output into a +4dBm input causes the opposite problem: The signal level is too low, and boosting it will likely increase any background noise.

For the most part, you shouldn't have to worry too much about theory. You should be able to plug equipment with ¼" jacks into anything else that has ¼" jacks (with the exception of instruments that use passive pickups, as noted above). At most, a little tweaking of a mixer channel's input pad (also called a *trim control*) should be all that's needed.

Cables and Connectors. Pro-level +4dBm signals are often carried on three-conductor XLR cables, which are used in *balanced* audio systems. Balanced cables resist noise build-up much better than two-conductor *unbalanced* cables, the kind that have ¼" phone plugs or RCA plugs on the end. Figures 3 and 4 show the difference between the two. (Some balanced cables do use ¼" plugs; in this case, the three conductors are wired to the tip, ring, and sleeve of a *stereo* plug.) These "balanced TRS" connectors carry only one channel of audio, even though they use a "stereo" plug. As you can probably guess, "TRS" stands for "tip-ring-sleeve.")

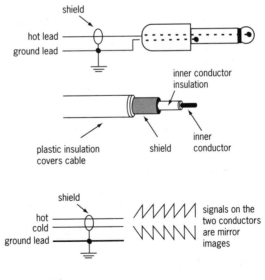

❸ In an unbalanced cable, the inner "hot" conductor is surrounded by a grounded shield that fences out hum and interference (although shielding is not 100% effective).

❹ In a balanced line system, the hot and cold leads carry the same signal, but are out of phase with one another.

❺ *Signals with opposite polarity cancel when summed together (a). In a balanced audio system, two "hot" connectors carry identical audio signals—but the polarity of one signal is reversed as it is being transmitted. At the receiving end, the polarity is flipped again, and the two signals are added (b). Any interference that was introduced to the signal as it traveled down the cable is now "out of phase," and will be canceled out, leaving only the desired audio signal (c).*

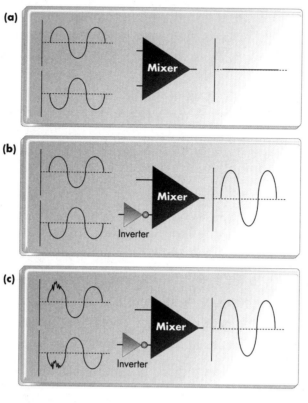

The reason balanced cables are less noisy is because the output signal going through the cable splits between two wires, a hot (+) signal and a "cold" or neutral (–) signal, which are reversed in phase. Both wires pick up noise as they travel down the cable, but on the receiving end, the phase-reversed signal is switched back in phase. So any noise is now in phase in one wire and out of phase in the other. Mixing these two back together causes the out-of-phase noise to cancel out the inphase noise, and the signal is (at least in theory) noise-free (see Figure 5).

Even unbalanced connections normally use *shielded* cable, which means that the thin metal strands that make the ground connection wrap around a layer of insulation surrounding the center "hot" wire. The shielding provides some protection against noise buildup. Unshielded lamp cord (also called "zip cord") is suitable *only* for audio connections between a power amp and speakers. Here the level is so high that noise buildup isn't likely to be a problem, but the ability to carry large amounts of current is crucial.

For the best audio performance, use the best cables and connectors you can afford, keep cable runs as short as possible, and avoid routing audio cables too close to power cables. An audio cable functions as a radio antenna; coiling a too-long cable magnifies the antenna effect. If you can't avoid placing audio cables in proximity to power cables, make sure they cross at an angle rather than run in parallel.

AC Plugs. What about power connectors? Most modern devices use three-prong, grounded plugs. Do not attempt to defeat the grounding by using three-to-two plug adapters. For best results, plug your entire music studio into a junction box that is hooked to a single outlet (assuming, of course, that it can handle the current your studio draws). This provides your entire studio with a common ground, minimizing the likelihood of *ground loops*, which are notorious sources of 60-cycle hum. If you're not sure whether the outlet you'll be using provides a true ground on its third prong, you can buy a tester at hardware stores for around $6. If your wall outlet doesn't provide a true ground, consult an electrician. Ungrounded electrical equipment is a safety hazard. (See Chapter 7, "Hums and Ground Loops," for more on grounding.)

Digital Audio

Digital audio is definitely a great leap forward for musicians wanting to make good recordings on a budget. The advantages of digital recording compared to analog (tape) recording include better potential signal-to-noise ratio (i.e., recordings with less "hiss"), cut-and-paste editing with no need to resort to a razor blade and tape splicing block, seamless integration between sequencers and audio recorders, and the ability to make perfect backup copies (clones). Among the disadvantages: Digital is less forgiving than analog tape of a too-hot input signal.

A/D and D/A Converters. When recording an analog signal (such as a microphone, guitar amp output, etc.) into a digital system, the signal first enters an analog-to-digital converter (ADC or A/D). This reads the voltage of the incoming analog signal, and converts it into the binary equivalent (in other words, the A/D converter expresses the voltage numerically as a series of ones and zeroes, which is the only language that computers understand).

When the digital data plays back, the ones and zeros pass through a digital-to-analog converter (DAC or D/A), which translates the numbers back into voltages. These voltages, which are essentially the same as those coming from the line output of a cassette deck or CD player, can proceed to an amplifier for monitoring.

Sampling Rate. The process of capturing the incoming signal's voltage is called *sampling*, and the number of times per second that the sampling process occurs is the *sampling rate*. Generally, a higher sampling rate more accurately represents high frequency waveforms. A rate of 44,100 times per second, as used on consumer CDs, is called a 44.1kHz sampling rate. Other sampling rates commonly found on digital audio systems include 96kHz, 88kHz, 48kHz, 32kHz, 22.05kHz, and 11.025kHz. 96kHz is the ultra-hi-fi rate specified for DVD; 22.05kHz and 11.025kHz sound more like AM radio.

Resolution. Typically, each sample is stored as a 16-bit *word*—a string of 16 ones and zeroes. The number of bits corresponds to the number of discrete digital numbers available to represent the incoming analog voltages, which specifies the signal's *resolution*. For example, a one-bit piece of data can represent only two values—on and off. Two bits can indicate four possible values (both on, both off, 1st bit on and 2nd bit off, or 1st bit off and 2nd bit on). Four bits can give 16 values, five bits 32 values, and so on.

The bottom line is that more bits allow more accurate quantifying of specific voltages, just as increasing the number of lines per inch with video gives a higher-definition picture. And the more discrete voltages you can measure (65,536 in the case of a

16-bit recorder), the greater the system's dynamic range and signal-to-noise ratio. This means that the softs are softer, the louds are louder, and the numbers that represent the stored sound more accurately represent the original signal (see Figure 6).

For this reason, more and more gear is going to 20-bit and even 24-bit resolution. While this requires storing 25% and 50% more data respectively than a 16-bit signal, the difference in sound quality is significant.

❻ *Sound is digitalized by rapidly measuring its level and storing the measurements. Two primary factors affect digital sound quality: the frequency of the measurements (sampling rate) and their resolution (bit depth). Above, you can see the individual measurements. If we decreased the sampling rate, the vertical spikes would start to disappear, obscuring the shape of the sound, and dulling it. Similarly, the lower the bit depth, the less accurate the level measurements can be. An 8-bit converter will divide the level scale into 256 possible values, whereas a 16-bit one will allow 65,536.*

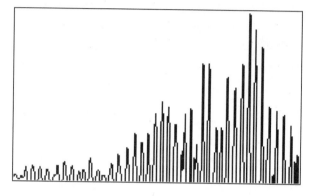

Quantization Noise and Aliasing. On paper, 16-bit "encoding" combined with a 44.1kHz sample rate equals high-fidelity, CD-quality audio. But in practice, other factors—particularly the hardware used in the system—affect the overall sonic performance. Not all 16-bit, 44.1kHz converters exhibit the same audio quality. The problem is not with digital audio *per se*, but with its communication with the analog world. Both analog-to-digital (A/D) and digital-to-analog (D/A) conversion can alter the sound, due to factors such as *quantization noise* and *aliasing*.

Quantization noise occurs due to differences between the digitized signal and the original analog signal. Aliasing occurs when a converter fails to grab both the positive and negative parts of a wave's cycle, resulting in an inaccurate representation of the original waveform (the converter records a lower frequency waveform instead). According to the *Nyquist Theorum*, the highest frequency you can record without aliasing is one half of the sampling rate. Therefore, if you record with a sampling rate of 22kHz, you can't capture a 12kHz frequency.

S/PDIF and AES/EBU. Two types of connectors are commonly used for stereo digital audio. The S/PDIF (Sony/Philips Digital Interface) format uses RCA (phono) or fiber-optic (TOSLINK) cables, while the AES/EBU (Audio Engineering Society/ European Broadcast Union) format generally uses XLR cables. Both formats carry left and right stereo signals together on a single cable. If you're planning to add a DAT deck to your computer music system, for example, ideally you'd like both the DAT deck and your sound card to have S/PDIF, AES/EBU, or optical input and output jacks, as this will allow transferring your music to and from the DAT in digital form, without passing through an intervening digital-to-analog and analog-to-digital conversion process. (This conversion process always degrades the audio, though with good converters the impact will be minimal.)

Digital Audio Recorder Overview

Digital audio recorders come in several different forms: multitrack tape decks, stand-alone hard disk recorders, portable MiniDisc multitrack recorders, and computer-based hard disk recorders. (Computer-based recorders will be covered in the section "Computers and Music.") DAT (digital audio tape) decks and CD recorders (the newest addition to digital audio recording) are two-track recorders used mostly for "mixdown"—tracks from multitrack recorders are combined onto these recorders to provide a final, two-track master suitable for mass-producing CDs and cassettes. They can also make backup, or archive, copies.

For more information on some popular multitrack digital recorders, see Chapter 2, "What Do I Need?"

Multitrack Digital Tape Recorders. Multitrack digital tape decks currently come in two main families based on the Alesis ADAT and TASCAM DA-88. The two use different tape formats (S-VHS and Hi-8 respectively), so tapes recorded on one can't play back on the other. Both systems provide eight audio tracks, but locking together multiple eight-track recorders expands the track count, eight tracks at a time. This explains why these machines are sometimes called *modular digital multi-tracks* (MDM), a term coined by *Mix* editor George Petersen.

Stand-Alone Hard Disk Recorders. These tapeless digital recorders store their audio on computer-type hard disks. They operate much like an ADAT or DA-88—or, for that matter, an analog multitrack—complete with front-panel level meters for each track, and rear-panel audio inputs and outputs.

The enormous advantage of tapeless recording is *random access.* Magnetic tape, either analog or digital, records sound linearly from beginning to end. For example, the sound at 1'00" is physically separated from the sound at 1'30" by a length of tape. Playing the two segments of audio back-to-back requires a razor blade and splicing block. (Actually, this primitive method of analog tape splicing *can't* be used with most digital tape, since it would disrupt codes that are recorded onto the tape and because most digital tape uses a rotating head that creates diagonal tracks across the tape.) In contrast, hard disk systems can "rewind" or "fast-forward" instantly from any point to any other. Most of them allow widely separated segments of audio data (or "tracks") to play back seamlessly, in any order.

Portable Digital Multitrack (MiniDisc) Recorders. MiniDisc recorders resemble analog cassette-based four-track personal studios, yet boast superior sound quality. They use MD data discs for audio storage of four or eight tracks, depending on the model (see Chapter 2, "MiniDisc Recorders").

DAT Decks. DATs are the most basic digital tape recorders. These machines offer two tracks of digital audio—more precisely, *stereo* digital audio. (Distinguishing between "two-track" and "stereo" is necessary because most DAT recorders do not allow recording of the left and right tracks independently; both tracks must be recorded at the same time.) The primary application for DAT in modern recording is creating a *master tape.* If you're happy with your recording setup and simply want a high-quality recorder that will allow creating a clean, quiet stereo master suitable for transferring to CD, DAT's the ticket.

There are other reasons for considering a DAT recorder: They can back up digital audio data, and the converters they contain can provide the audio in/out for hard disk recording systems. Their downside: They were never really designed as a "pro" medium, so DAT cassettes are relatively fragile.

DAT cassettes, which look like miniature VCR tapes, can record stereo audio with slightly higher fidelity than CDs (a 48kHz instead of 44.1kHz sampling rate). However, most musicians record at 44.1kHz, even with DAT, to avoid having to convert the sample rate from 48kHz to the 44.1kHz rate used by CDs (the sample rate conversion process can color the sound somewhat).

CD Recorders. CD recorders (also called CD-R drives or CD writers) used to cost tens of thousands of dollars, but in the past five years their prices have fallen dramatically, giving the home studio musician the ability to record and make one-at-a-time CDs without having to work with a mastering and duplicating facility. Rewritable CDs allow you to rewrite information that has previously been recorded. Some CD recording software also gives you the ability to create photo CDs and video CDs. See Chapter 2, "CD Recorders."

MIDI

You can record digital audio and never touch a computer, but why limit yourself? The advent of MIDI in 1983 made it practical to hook up low-cost personal computers to synthesizers, and musicians' lives have never been the same since.

But first, here's a little history. By the early '80s, several manufacturers were building synthesizers that were controlled internally by digital messages. The messages told the built-in microprocessor which keys were being struck on the keyboard and which knobs were being turned; this data could even be recorded on the digital sequencers that were being built at the time. But each manufacturer had a different method of encoding the messages, so a Roland sequencer could only work with a Roland synth, a Sequential Circuits sequencer with a Sequential synth, and so on.

In a rare, and highly laudable, instance of industry-wide cooperation, many companies banded together to produce the Musical Instrument Digital Interface (MIDI), a universal form of communication that allowed synths and sequencers built by different manufacturers to talk to one another. Debuting in 1983, MIDI kicked off an unprecedented period of growth in the electronic music industry. Other factors, such as the declining cost of microprocessors, played key roles too, but MIDI was the glue that held it all together.

Originally, MIDI was designed to handle a fairly simple set of tasks. The initial idea was that when you played one MIDI synthesizer's keyboard and attached a cable from its MIDI out to another synthesizer's MIDI in, the second synth would receive messages from the first synth that told it which notes to play, thus layering the sounds of the two devices. The most basic messages are *note-on* and *note-off*; each MIDI note-on also includes a *key velocity* message, which tells the receiving device how hard the key is being struck. Other performance data, such as pitch-bends and modulation wheel moves, could also be transmitted as MIDI data. Timing synchronization between devices such as drum machines and sequencers was handled by MIDI clock messages (see Chapter 4 for more on synchronization).

Over the years, MIDI has been expanded to allow for more complex types of communication, but the original MIDI specification has never been abandoned. While MIDI instruments built in 1983 don't have some of the advanced functionality that has been grafted onto the MIDI spec, they still work with MIDI gear built today.

MIDI remains popular for sequencing, live control of synths (and samplers) from master keyboards, and studio control (typically, signal processors and mixers). It's also used for sending patch data from synths to storage devices, and synchronization via MIDI time code.

MIDI is not without limitations. Its bandwidth (31.25 kbaud) is just barely high enough to handle multitrack music performances without choking. The original implementation called for only 16 separate channels of music data on one cable, which seemed plenty at the time—but today, rigs that require 32 or more channels are quite common. (This requires the use of *multiport* MIDI interfaces for computers, as described later.) Most importantly, MIDI is a keyboard-oriented language, and doesn't work well to describe performances on instruments like electric violin. Alternatives and extensions have been proposed over the years and added to the spec, thus keeping it vital well over a decade and a half after its introduction. In any event, the huge installed base of MIDI devices makes it extremely unlikely that MIDI will become obsolete any time soon.

MIDI Interfaces and Operating Systems. Getting MIDI in and out of a computer requires a specialized piece of hardware called a *MIDI interface.* Except for Atari ST-series computers and the now-obsolete Yamaha C8, both of which had built-in MIDI jacks, the MIDI interface is a separate piece of hardware. On the Macintosh, this plugs into the printer or modem port (and sometimes both if you need lots of MIDI channels). Serial and parallel port interfaces exist for the PC. Most sound cards have MIDI in/out capability, but require an adapter since MIDI jacks are slightly too wide to fit in a standard PC's rear-panel slot. The adapter usually has two or more MIDI jacks (or plugs) connecting to a DB-15 connector that plugs into a matching connector on the sound card. Some sound cards include this adapter, while other cards require purchasing the adapter as an accessory.

Note that more and more pro-level MIDI interfaces are *dual-platform*, meaning that they can work with both Windows and Mac-family computers. Also, these higher-end devices also provide more than one MIDI output and input. For example, there may be four MIDI output jacks, allowing the device to address 64 separate MIDI channels (16 MIDI channels x 4 cables = 64 channels total). In many instances, these interfaces can even be stacked to provide more channels. Do you really need, for example, 256 MIDI channels? Probably not, but it means you can hook up a large system to your computer without having to do any re-patching.

In addition to the interface, computers need some sort of MIDI operating system to route data from the application (the sequencer or editor/librarian, for example) to and from the interface. In the PC, Windows MIDI drivers can be installed either from the Windows master diskettes/CD or from a disk provided by the interface manufacturer. On the Macintosh side, some music software uses Apple's cumbersome MIDI Manager system extension, but thankfully, this has been largely superseded by Opcode's OMS (Open MIDI System) and to a lesser extent, Mark of the Unicorn's FreeMIDI. These programs recognize your studio setup (the MIDI devices connected to your interface), and you can select the appropriate model names (K2000, ZR-76, etc.) from a menu instead of referring to them as something like "the device connected to cable A." The Name Manager in OMS also learns all of the patch names in your synths by retrieving them from OMS-compatible librarian documents. Thus when you choose a record track in a program like Studio Vision, you can choose device and sound assignments by name.

In addition to routing MIDI data, MIDI operating systems can provide real-time MIDI processing (e.g., keyboard splitting, delay, transposition, etc.) and also make information, such as the names of the patches currently being used on particular synthesizers, available to other MIDI applications.

Computers and Music

Computers can record and play back MIDI data, edit and print out music notation, customize and store synthesizer patches, and record, play back, and edit digital audio. We'll look at each application in turn.

Sequencers. Sequencers record the MIDI events that are generated as you play
MIDI-compatible instruments, and play those events back into MIDI-compatible gear
to re-create your performance. Because MIDI consists of data, not actual sound, the
sound depends on what plays back the MIDI data. MIDI is conceptually like the
holes in a player piano's paper roll; if you put the paper roll in a piano that's out of
tune, or one in which certain strings are missing, the sound will be altered even
though the paper roll doesn't change. Sequencers also allow editing of the MIDI data
after it's been recorded, so you can fix bad notes, timing problems, etc.

When a sequencer sends commands to the receiving module in the form of MIDI
messages, the sequencer doesn't care how (or even whether) the module responds. It
could be playing notes, or turning light bulbs on and off.

Think of a sequencer as the musical equivalent of a word processor. Like a word
processor, you can cut and paste blocks of data (in this case, MIDI notes rather than
alphabetical characters; see Figure 7. You can alter the data in numerous ways, just
the way you can change the wording of a sentence in a word processor, then save
your work to disk.

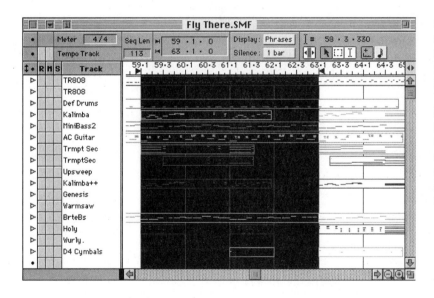

❼ Clicking and dragging
across a four-bar region (in
this example, from measure
59, beat 1, to measure 63,
beat 1) in Opcode Vision's
Track Overview window highlights that area. It's now
ready to be cut, copied, or
otherwise processed. The
rectangular boxes in the window show where MIDI data
has been recorded on each
track.

A sequencer is more complex than the average word processor, however, because
MIDI data has more dimensions than text. For each note in a sequence, the sequencer
needs to store, and allow you to edit, at least the following note-related parameters:

- Pitch (also called the note number)
- Velocity (how hard you struck the key, which gives a value for dynamics)
- MIDI channel
- Start time
- Duration (note length)

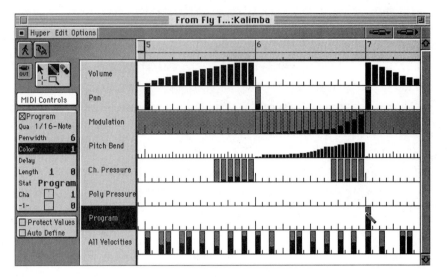

8 *Emagic Logic can show several strips of continuous controllers simultaneously in the HyperEdit window. The strips can have their own background to make them easier to differentiate. The pencil tool is entering a program change command at the beginning of measure 7.*

MIDI sequencers also record and play back many types of events other than notes. Pitch-bends, controller data (vibrato, timbral changes, etc.), and program changes to select different instrument sounds are the most common of these events (see Figure 8 and 9).

Like a multitrack tape deck, sequencers organize their music data into *tracks.* Each track will typically (though not necessarily) be dedicated to a single MIDI channel of notes, such as the piano or bass part. Tracks can be individually muted so that you can hear the rest of the music without the muted track. With *punch-in recording,* you can replace part of a single track—for instance, bars 9 through 11—and leave the rest of the music untouched.

One common sequencer editing process, *quantization* (sometimes called auto-correct), tightens up a performance's timing by shifting each note forward or backward slightly so that its start lines up with the nearest specified rhythmic value. The user defines the quantization "grid" to which notes will be adjusted, such as 16th or triplets.

In order to do this job, or any other related to rhythm, a sequencer must have an internal clock to keep track of the music's timing. The more precise the clock, the more accurately it will record and play back music. A sequencer clock's precision is measured in ppq (pulses per quarter-note). A basic sequencer may offer only 24 ppq of clock resolution, but this isn't enough to represent the subtleties of a musician's performance. The clock resolutions on typical sequencers range from 96 to 960 ppq. However, note that finer resolutions may be irrelevant if the computer on which the sequencer is running has "loose" timing (generally caused by a slow processor, or multi-tasking operation). For more on quantization, see Chapter 4.

Many sequencers also contain software "mixers," where the mouse can control a screen full of virtual faders and buttons. These controllers send MIDI data that adjusts the volume, stereo panning, or other aspects of the synth(s) being controlled

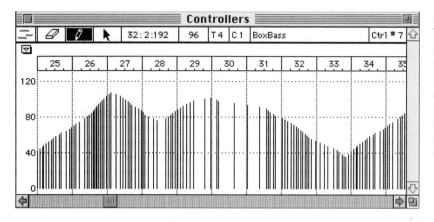

9 *The Controllers window from Passport Pro 6. It's currently set to display Controller #7 (MIDI volume), as shown in the upper right box; therefore, this curve is controlling the main volume for the selected channel (in this case, channel 1, as indicated to the left of the track-name box).*

by the MIDI messages (see Figure 10). Sequencers can also automate non-MIDI, audio tracks via MIDI-controlled volume faders. The fader inserts in an analog signal path; the fader level depends on an incoming MIDI controller value. Using different MIDI controllers, or sending data over different channels, allows automating multiple faders.

With sequencer software and a MIDI interface (or MIDI-savvy sound card), a computer can record and play back arrangements using anything from a single, multi-timbral synthesizer (i.e., a synth capable of playing several sounds at once) all the way up to a fully loaded rack of synths. Sequencing a song allows building up layer upon layer of music, with the ability to edit and revise as you go along. Various types of graphic displays of the music data, from standard music notation to numeric lists of the recorded data, help streamline this editing process.

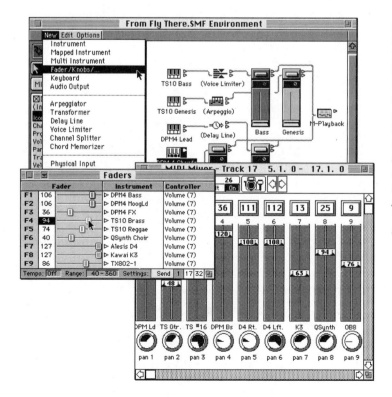

10 *Three different approaches to faders. The top window, from Logic, shows not just faders, but other processors (arpeggiator, delay line, etc.) that you can insert into your "virtual mixer." At this moment, a fader is about to be created on screen using the pull-down menu. The lower left window shows Vision's easily understood fader implementation. At lower right is a Cubase "mixermap," set up for program changes, mixing, and panning.*

Some programs integrate MIDI and digital audio hard disk recording, so you can record MIDI and audio data in the same environment. Or, a sequencer can sync to digital tape or hard disk, and send data to MIDI instruments while the digital recording system handles the audio.

Notation Software. Many software sequencers offer some form of traditional staff notation (see Figures 11 and 12), and almost all will transcribe in real time. You can edit the sequence by dragging notes around on the screen, and then print out the results. These notation utilities are fine for many purposes, including songwriter lead sheets (sheet music with melody, lyrics, and perhaps piano accompaniment). Dedicated notation software, however, typically provides more convenience features, more specialized symbols, and more precise control over the page's appearance. Notation software may also include some MIDI sequencing capabilities, and will generally allow you to enter note pitches by playing them on a MIDI keyboard.

Editor/Librarians. An editor/librarian program stores the bank of patches (sound programs) from your synthesizer's internal memory onto a computer's storage medium (floppy disk, hard disk, etc.). You can generally combine patches from separate banks into a single new bank, and edit the patches themselves using the computer screen's large, detailed graphics rather than the synth's small LCD. If the synth has been popular enough to garner third-party support, you can buy banks of patches from sound developers on a floppy disk or CD-ROM (or download them from an online service) and transmit the data to your synth. A sophisticated editor/librarian includes some database features, allowing you to search for patches by category or keyword. For instance, you could ask the program to find all bass patches with "fretless" as a keyword.

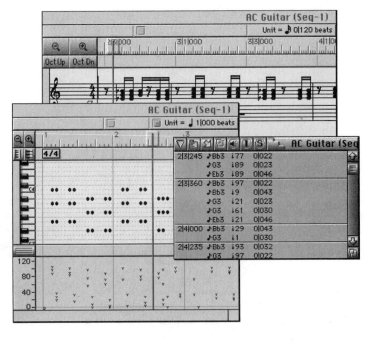

⓫ Mark of the Unicorn's Performer sequencer offers notation editing (upper window), along with piano roll editing (lower left window) and event list editing (foreground window). Each type of editing window has associated advantages and disadvantages: notation is the most familiar to many musicians, piano roll editing provides lots of information at a glance, and event list editing allows for extremely precise, detailed editing.

12 *The screen shot of Opcode Vision (Mac, PC) shows some of the features found on most software MIDI sequencers. At the top are the transport controls, which emulate the controls on a tape deck. The next window down shows a track overview; if you look closely, you can see the individual notes recorded in each track, expressed as bars of color. The window at the bottom right is close-up of a single track; the vertical bars represent note velocities. The track can also be viewed in standard notation. Not shown is the list view window, which expresses each MIDI event in text form.*

Sequencer/Recorders. As mentioned earlier, some programs combine sequencing with digital audio recording. Programs like Opcode's *Studio Vision*, Mark of the Unicorn's *Digital Performer*, Cakewalk Software's *Cakewalk Pro Audio*, Steinberg's *Cubase Audio*, and Emagic's *Logic Audio* treat audio and MIDI data similarly, allowing you to assemble a song that includes both synthesizer parts and acoustic tracks in a single integrated environment. Simple audio recording software may record and play back only one track of audio, either mono or stereo. Professional software often has 16 tracks or more (see Figure 13), but higher track counts generally require a computer with a very fast processor, and a decent amount of RAM.

You can also use the sync output from an external multitrack tape recorder to drive a MIDI sequencer program, which is another way to synchronize audio and

13 *This screen shot from Digidesign's Pro Tools recording system shows five tracks of digital audio and two of MIDI data. Note that mixing functions such as levels and panning can be handled by the computer rather than an external hardware mixer.*

MIDI. However, since these decks typically have eight audio outputs, you'll need a mixer that has a sufficient number of inputs. Another option is to run a hard disk recorder *concurrently* with a MIDI sequencer. In other words, two programs run simultaneously within the computer, and are synchronized to a common timing signal (on the Mac, this is usually generated by OMS, a MIDI-oriented operating system extension).

Sound Cards. Audio recording software requires a computer with analog-to-digital (A/D) and digital-to-analog (D/A) converters, typically in the form of a plug-in card. Without such converters, there is no convenient way to get the audio in and out of the computer. One exception is Macintosh AV-series computers and PowerMacs, which include internal converters. (Earlier Macintoshes, from the LC on, had audio I/O but it was converted to and from eight bits, which is very low resolution compared to the 16-bit resolution used by CDs.) Note that not all converters offer the same sound quality. Pro recordists use cards with higher-quality converters designed specifically for pro audio.

If all you need is stereo recording, the built-in digital audio converters on the new Macs are adequate. Depending on the software, a Mac can play back four or more channels of sound, but the audio will be mixed to two channels (in other words, to stereo) before being sent to the hardware output jacks. However, there are sound cards for both Windows and Mac machines that offer more than two channels of simultaneous audio recording and playback. See "Choosing a Sound Card" in Chapter 2 for more information on sound cards.

Effects Plug-Ins. An audio plug-in is a piece of software that provides additional functionality to a "host" program. For example, the plug-in might provide guitar distortion effects within a hard disk recording program that otherwise offers only basic processing, such as tone and dynamics control. Plug-ins often replace effects that used to be done in hardware, such as compression, expansion, reverb, delay, EQ, gating, noise reduction, vocoding, spectrum analysis, surround sound, pitch correction, analog tape simulation, tube emulation, and a lot more.

Plug-in architectures are not all compatible, however, so a plug-in that works with one program may not work with another. For Windows, the DirectX plug-in format has become a de facto standard. On the Mac, there are several popular formats including Digidesign's TDM and Sound Designer format, Steinberg's VST, Adobe Premiere, and others. With a fast computer and efficient code, plug-ins may provide "real time" control, where changing a parameter affects the sound almost instantly. Otherwise, you'll need to wait while the plug-in processes the file before you can hear the results.

Data Storage. Recorded audio is stored on a computer's hard disk (either the internal disk or an external disk). If you do the math (2 bytes x 2 channels x 44,100 x 60 seconds), one minute of stereo audio sampled in a 16-bit format at 44.1kHz takes up more than 10MB (megabytes) of hard disk space. That's one reason why tapeless recording is not cheap. However, as hard drive prices (especially cartridge-based, removable types) have plummeted, it has become a more attractive option.

Hard Drives. Hard disk drives are sometimes internal—i.e., mounted inside the computer or sampler—and sometimes external. The most common way of linking external hard drives to the host device is with a SCSI cable. SCSI (the Small Computer Systems Interface, pronounced "scuzzy") is a communications protocol that allows up to eight devices, including hard drives and CD-ROM drives, to share data. SCSI connections are not quite as simple or trouble-free as MIDI connections. Cables must be short, cannot be plugged or unplugged while power is on, and use several different connector types (bring on the adapters!). SCSI is built into Macs, but Windows machines typically require a SCSI adapter card, such as those made by Adaptec. SCSI cards are fairly common for Windows machines, so your nearby computer store is likely to have a decent selection, as well as someone who can help you choose the right type for your application.

Drives used for digital audio require a fast access time (better than 15 ms, although that's sluggish if you expect to get lots of tracks), high rotational speed (7200 rpm is typical), and "A/V" operation. Hard disks periodically calibrate themselves, which you don't want to have happen while recording digital audio or a CD. A/V types postpone calibration to periods of inactivity. In any event, the faster the drive, the better the performance; interestingly, bigger drives are often faster than smaller-capacity drives.

Computer hard disks come in sizes from 40MB up to 9 gigabytes (billions of bytes, abbreviated GB) and more. At present hard drives are rapidly growing in capacity and speed (in fact, it's getting difficult to buy a drive smaller than 1GB), and prices continue to drop.

Removable-Media Drives. Removable-media drives are also upping their speed and capacity. Early drives featured 44 or 88MB of storage; 100MB disks (e.g., the Iomega ZIP format) are now commonplace, and 1 and 2GB drives (such as Iomega's Jaz drive, currently a popular format for digital audio) have established themselves as yet another standard. Iomega is not the only game in town; SyQuest, one of the first names in removable drives (see Figure 14), currently offers the EZFlyer (230MB) and the SyJet (1.5GB) drives.

14 *Removable hard disks are an ideal way to store digital samples, which are often too big to fit on floppies. The SyQuest cartridge is a popular removable storage medium. This 200MB model is probably the last in the 5-inch format; newer models use a 3-inch cartridge.*

A Jaz drive is not the same thing as a hard drive, though you may be tempted to treat it similarly. (See box below for tips on Jaz drive maintenance.) Before defragmenting a Jaz drive, for instance, you should check with the manufacturer's technical support, as Norton and other disk optimizers can damage the drive's disk blocks. Jaz drives are also vulnerable to strong magnetic fields, particularly coming from computer monitors, so keep these drives (as well as Zip drives) about three or four feet away from your monitor.

Recording to removable cartridges is doable, but in general, these are best for backing up audio. It's more common to record audio to a traditional hard drive, then back up the data to a removable drive. Other removable backup options include *magneto-optical* drives (which, while slower than fixed hard drives, are very robust and can store up to 4.6GB of data) and special DAT cassette decks (these store 2.4GB and more with data compression; ordinary DAT recorders can error-correct digital audio without perceptible signal degradation, but they're not reliable for computer files, because individual bytes can become corrupted). However, backing up to DAT is quite slow.

Recordable CDs are another option, as the prices of both CD recording hardware and blank CDs have declined dramatically in recent years (soon CD recorders may become as common in home studios as DAT decks, if not more so). CD-recordable media have a storage capacity of 74 minutes. Rewritable CDs have also recently come on the market, and are more environmentally friendly than the non-reusable, "write-once" types.

TIPS TO HELP JAZ RELIABILITY:

1. Keep Jaz drives oriented so that the cartridges are horizontal. Do not place them on edge with cartridges vertical.

2. Avoid Jaz drives that lack adequate cooling. Iomega-labeled external drives lack cooling fans, which are typically used in third-party external drives.

3. Avoid using Jaz drives in servers or other applications that place the drive and cartridge in constant use.

4. If problems occur, allow drive and cartridge to cool. You may then be able to extract a troublesome cartridge or read its data.

5. If data cannot be read from a cartridge, Iomega may be able to recover the data for you. The two layers of a cartridge reportedly can shift, particularly with excess heat, requiring re-alignment.

6. Try enabling write-verification in the Jaz drivers. This may help data integrity substantially.

7. Avoid placing Jaz drives near magnetic fields, such as those created next to monitors.

MiniDiscs are a long-shot data storage medium. Although they have achieved some success as a musical playback and recording medium (as of this writing, their popularity is slowly increasing), it was expected on their introduction that they would also become a sort of "super-floppy" disk. That has not come to pass, and probably never will.

Synthesizers

Classic analog synthesizers use discrete analog components (transistors, diodes, capacitors, etc.) to create devices called *oscillators*. These generate simple, periodic electrical waveforms, such as sawtooth, triangle, and square waves (so named because of their shape when viewed on an oscilloscope; see Figure 15).

The oscillator's signal (or a mix of multiple oscillators) then goes through a *voltage-controlled filter* (VCF), which modifies the oscillator's basic timbre by changing its harmonic structure. Like most filters, this changes frequency response in a selectable, predictable way. The key filter parameter, *cutoff frequency*, determines which frequencies are boosted or attenuated, as opposed to passing through the filter unaltered.

The filtered signal then proceeds though a *voltage-controlled amplifier* (VCA), which determines the loudness contour (dynamics).

Both the VCF and VCA are controlled by modules called *envelope generators*, which shape the overall sound dynamically by changing the VCF cutoff frequency and VCA gain over the course of a note. With the help of the envelope generators, the raw buzz of an oscillator can swell and fade away slowly like a string orchestra, start and stop abruptly like an organ, or start quickly and fade out smoothly like a plucked guitar string.

Today, most synthesizers create sound digitally, but many of them still use the same concepts to shape sound. Instead of producing a simple periodic waveform, digital oscillators typically play back a digitally sampled recording of an actual sound, such as a note played on trumpet or bass guitar. This is called *sample playback* synthesis. Before reaching the output, the sample still passes through a digital filter, which shapes the tone, and through an amplifier stage, which applies an amplitude envelope to each note. You'll see lots of bells and whistles grafted onto the sample playback process, but the basics don't change much from one instrument to another.

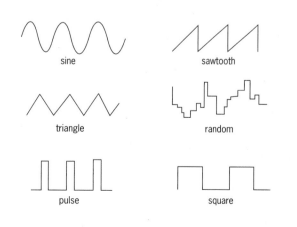

15 *A few cycles of classic synthesizer waveforms.*

sine

sawtooth

triangle

random

pulse

square

Sample playback is common in both $100 sound cards and $5,000 synthesizers, but many factors influence differences in sound quality between low-end instruments and expensive ones. The amount of memory available for samples (measured in megabytes of ROM) has much to do with the samples' variety and quality. The sample rate at which the samples were originally recorded, and the rate at which they're played back, affects the synth's high-frequency response. (Pro instruments typically have a 44.1kHz sample playback rate, the same rate as CD recordings, or higher.) More expensive instruments generally have better digital filters as well.

Transposition is an important part of sampling technology. To make a long story very short, a sample is recorded at one pitch—let's say a trumpet note at Middle C. Rather than record a separate sample for every key, which takes up a great deal of memory, playing notes above and below Middle C triggers the same sample. However, the keyboard digitally retunes the sample "on the fly" to a higher or lower pitch. Usually, a sample can survive being tuned up or down by a few half-steps without any drastic effect on the sound quality, but greater transpositions can introduce unwanted sonic artifacts. A sample that has been pitch-shifted too far down can sound grainy or dull; shifting too far up can sound thin and toylike. Then again, sometimes this is what you want.

To make a synth "trumpet" (or any other instrument) sound more realistic across a wide range of the keyboard, samples are taken at multiple pitches (this is called *multisampling*). These samples are then "mapped" to a keyboard so that each of them has to play only a narrow range of keys. For example, an instrument might be sampled every four or five semitones.

It's quite difficult to create a seamless timbral match when transitioning from one sample to another, even if the samples are of the same instrument, played by the same player, into the same microphone. This is especially true when one of the samples is being transposed upward (say, from Middle C to the F above it) and another downward (say, from the C an octave higher down to F#). The point on the keyboard at which one sample meets another is called a *multisample split point*. It takes considerable effort to avoid an undesirable jump in sound quality between one sample and the next. The number of samples in a multisample, the care with which they are matched, the lengths of the individual samples (longer is better), the creative use of filtering to "mellow out" pitches that are transposed upward, and the care with which the samples are looped are among the factors that affect the sound quality.

Other types of digital synthesis are not based on sample playback, but instead create complex, musically interesting timbres using mathematical formulas rather than recordings of "real" instruments. One familiar example is FM (frequency modulation) synthesis. Although FM has gotten a bad name because a simple form of it is used in inexpensive sound cards, in more powerful FM instruments it's capable of creating sounds that are both musically pleasing and more responsive to the player's musical gestures than sample playback. A newer technology, called *physical modeling*, goes much further than FM in generating expressive, responsive tones; it creates a computer model of an instrument's sound, and "plugs in" different numbers into these models to create the various tones.

Some electronic keyboard instruments only play preset sounds, while others are *user-programmable.* Synthesizer programming is the process of giving the sound-generating circuitry instructions about what it's supposed to do when you play a note. Programmable synths do this by allowing you to edit the values of certain *parameters* that influence sound quality, such as vibrato speed, filter cutoff frequency, oscillator tuning, attack time, and so on. Programmable synthesizers have a block of RAM (user memory) set aside for storing new sound programs, which may consist of hundreds of possible parameters.

Incidentally, sound programs are sometimes called *patches* rather than programs because 30 years ago, analog synthesizers used actual patch cords for connecting the oscillators, filters, and VCAs to one another.

Tone Modules. Once you can send note-on and note-off messages down a MIDI cable, there's no reason to put a set of keys on every sound generator. In fact, you can save a lot of money by buying one MIDI instrument with a keyboard, or for that matter a MIDI guitar or percussion controller (see "Alternate Controllers," below), and then adding to your rig by hooking up tone modules. (On the other hand, for live playing, you might want to have several keyboards handy.) A tone module typically has everything that a keyboard synthesizer has, except the keyboard, and makes sound only in response to incoming MIDI messages. (See Figure 16 for an example of a piano tone module.) As we explain a few more key concepts in this section, remember that these apply equally to synths and tone modules.

Most current tone modules are *multitimbral,* meaning that a single module can respond on several MIDI channels at once—usually on all 16 channels, but some multitimbral instruments are limited to fewer channels. Each channel will usually produce a different sound—perhaps piano on channel 1, strings on channel 2, bass

16 *Ever seen a piano without a keyboard? E-mu's Proformance Plus piano module contains stereo digital samples (recordings) of individual piano notes. Connect it to a keyboard's MIDI Out jack and you can play a virtual 9-foot grand. While the Proformance is optimized for piano sounds, many tone modules are designed to play 16 or more different sounds simultaneously, so you don't need to buy multiple boxes to create an ensemble piece.*

on channel 3, and so on. Each multitimbral part can play *polyphonically*—that is, play chords, up to the polyphonic voice limit of the module.

This is an important concept to grasp: Each tone module has a limit to how many voices (i.e., how many notes) it can play at once, which is called *polyphony*. Many modules provide 32-note polyphony or even more, but some of the richer, more interesting synth patches "layer" two or more voices per note. If you're using patches in which each note requires two voices, a "32-note polyphonic" module will only produce 16 notes simultaneously. If you try to play more notes than this, either the excess note-on messages will have no effect, or they will cause some notes that are still sounding to drop out. This phenomenon, sometimes called *note stealing*, can be especially troublesome when a single tone module plays back a complex musical arrangement.

Most tone modules come equipped with some sort of on-board effects processor (see "Effects Processors," below). This can add reverb, chorusing, or other sonic enhancements. You will probably be able to set the *effect send level* separately for each of the multitimbral parts. For example, you might want lots of reverb on the piano but only a little on the bass, and both reverb and chorus on the strings. Many tone modules have only one effects processor, so the available processor(s) must be shared by all of the multitimbral parts. However, more sophisticated modules allow individual processing for specific parts.

If you're using the module to play only one sound—that is, not playing multitimbrally—you should be able to obtain a richer, fuller sound by using the effects settings that have been programmed into the individual patch. But individual patches that were programmed with spectacular effects may sound a lot less wonderful when used in a multitimbral setup that has more generic effects. Conversely, if you use the module's copy utility to copy the great effect in the string patch to your multitimbral setup, the bass, piano, and drums may sound quite odd because they have to share the string effect. Since the module has only a limited number of hardware effects processors, it's often not possible to use several patches in the same multitimbral setup and have each patch retain its original effects.

A tone module's *operating system* is the on-board software that reads and responds to the front-panel buttons and displays messages in the LCD (liquid crystal display) or other display. Operating systems are sometimes upgraded by manufacturers to provide enhancements or fix software bugs. The upgrade will usually be in the form of a chip that plugs into a socket on the circuit board, although some samplers (see below) load their operating system from a floppy at the start of each session. If you're not familiar with how to install a chip, have it done by an authorized service center. While replacing a chip is easy, if you accidentally break off one of the chip's pins, plug it in backwards, or fry it with a static electricity discharge, the chip can't be fixed.

Alternate Controllers. As mentioned above, MIDI was developed as a language for controlling electronic keyboard instruments. However, not everyone wanted to abandon performance techniques perfected over years of practice on guitars, drums, and wind instruments in order to be able to make noise with MIDI tone modules. Not only that, but the traditional black-and-white keyboard is just plain lousy at certain types of expressive musical nuances. To address these two problems, a number of alternative MIDI controllers have been developed (see Figure 17).

🔊 *MIDI lets you separate the controlling instrument from the sound generator. This wind controller provides an alternative to the traditional piano-style keyboard. The strength of the player's breath can be tied to the volume of the note.*

Alternate controllers actually predate MIDI: The Lyricon wind synthesizer, for example, used a saxophone-type interface to actuate an analog synthesizer. Robert Moog also produced ribbon and drum controllers.

MIDI has made it practical for manufacturers of alternate controllers to do what they do best—build a sensitive performance interface—while taking advantage of existing tone-generator technology. For example, percussion controllers can hook up to any drum machine or sampler. More visionary alternate controllers like the Buchla Thunder (touch pads) and Buchla Lightning (infrared wands) have also appeared from time to time; guitar, Chapman Stick, violin, voice, and many other MIDI controllers are available (and achieve varying degrees of success, both in terms of commerciality and ability to translate performance gestures to MIDI).

Another option for players of traditional instruments is the *pitch-to-MIDI converter.* This translates a monophonic musical line, such as you might produce by singing or playing sax into a microphone, into a stream of MIDI data. Pitch-to-MIDI conversion technology is not perfect, but works surprisingly well, especially if you're using it as an input into a sequencer, where you can clean up any glitches that the converter might introduce.

MIDI Master Keyboards. MIDI master keyboards look like a keyboard synthesizer, but they only transmit MIDI data and make no sound. Small, lightweight master keyboards are especially useful with computer sound card music setups; some models even have a recess in the upper panel where you can set the computer's QWERTY keyboard.

Why not buy a master keyboard that also makes sound—in other words, a synthesizer—thus saving money by getting both keys and a sound generator at the same time? That's a valid question. However, to meet a particular price point, a synth might skimp a bit on keyboard features, whereas a dedicated master keyboard is optimized for playing "feel" and flexibility. It may have a better mechanical action—many have 88 keys that are weighted to simulate a piano action—as well as a bank of programmable buttons and sliders that can send out any MIDI message. You may also be able to program the keyboard to split into as many as eight separate *zones*, each transmitting on its own MIDI channel. This allows up to eight different sounds to play at once, or be layered with one another for a fuller sound.

Drum Machines. For a few short years between the advent of affordable microprocessors and MIDI's birth, drum machines were a vital part of pop music. They included both a selection of drum sounds (either synthesized simulations or actual digital recordings) and also a sequencer for recording drum beats—often called *patterns*—and stringing the patterns together into songs. In place of a keyboard, a drum machine offers a bank of *pads* to which you can assign various drum sounds (see Figure 18). In recent years, all but the least expensive drum machines have had velocity-sensitive pads. Hit them harder and the sound gets louder, just as it does on a real drum.

18 *A drum machine is a synthesizer that's been optimized for playing back percussion sounds. The 16 large buttons on this Akai MPC3000 can each play a different sound, which are typically drum samples. Drum machines also include pattern-based sequencers that allow you to build up drum beats by overdubbing new sounds on a looping track.*

MIDI sequencing has cut into the drum machine's turf pretty heavily. Most musicians prefer to record an entire sequence (drum parts and other parts) in a single sequencer. You'll still see dedicated percussion tone modules now and then, but they tend not to have built-in sequencing.

Another variation on the theme is the "groove box," which combines drum-machine-style pads with sampling and full-featured MIDI sequencing. Many players swear by these because of their fast, efficient interface which makes it easy to lay down grooves quickly.

One advantage of drum machines is that striking their pads with your fingers is more like drumming than entering a drum part on a keyboard. Some machines also have features like dedicated hi-hat open/close sliders, or the ability to add types of performance nuances that can't easily be duplicated by a MIDI sequencer. As a result, many rap and dance music artists use drum machines for their real-time features.

General MIDI

General MIDI is where synthesizers, MIDI, and consumer electronics meet. As MIDI technology began to appear in consumer-oriented products such as computer sound cards and digital pianos, a problem surfaced. Each manufacturer used their own method of storing patches in memory. Remember, MIDI doesn't specify actual sounds; it only transmits data. A MIDI program change message with a value of 37 might call up a flute program on one synth, a combo organ on a second synth, and a drum kit on a third. Especially when sharing Standard MIDI Files (see below) among MIDI equipment, getting the music to sound as the composer or arranger intended can be difficult or impossible without detailed notes that specify which instruments should correspond to which program changes.

General MIDI (GM) addresses this problem, and some related ones. GM is not a change in MIDI data format; rather, it's a list of basic features required by any synth that wants to claim GM compatibility (see "The General MIDI Rulebook" sidebar). First and foremost, GM provides a list of 128 standard sound programs. Program 37, for instance, is *always* a slap electric bass sound. An arranger who wants to be sure that a GM playback synth plays a slap electric bass for a particular track can simply insert a program change message with a value of 37 at the beginning of the track.

GM has some additional requirements. A GM instrument must provide at least 24 notes of polyphony, respond on all 16 MIDI channels at once, and its drum kit (which is always assigned to channel 10) must be laid out in a specific fashion—kick drum on low C, snare drum on D, and so on.

Note that GM does not impose an *upper* end on an instrument's capabilities. For example, it can have 96-note polyphony and 512 programs, and still qualify as a GM instrument. GM provides only a *minimum* standard for instrument performance. Both Roland and Yamaha have proposed *supersets* to GM. Roland's GS instruments and Yamaha's XG instruments comply with the GM spec, but they offer an added palette of features that are available to other synth manufacturers, should they wish to use them, and to developers of SMFs, who can put GS and/or XG data into their sequences to enhance the quality of the music (see "General Plus" sidebar).

Problems with GM. General MIDI was intended to solve problems, not create them. However, some GM definitions are rather fuzzy. Consider polyphony: What is a voice? GM says that an instrument must have a minimum of 24 voices. This sounds fine until you realize that some manufacturers refer, perfectly legitimately, to a voice as an instrument's simplest, individual sound-producing generator. In theory, an instrument can be 24-voice, but in practice, since all its sounds may use two or three

THE GENERAL MIDI RULEBOOK

In order to conform to GM System Level 1, a sound source must:

- Respond to messages on all 16 MIDI channels.
- Be 24-voice polyphonic, with dynamic allocation (polyphony can be 16-voice for pitched instruments, 8-voice for drums). This means that the synthesizer must be able to play 24 simultaneous voices (notes), and automatically reassign voices to play new notes whenever required by the musical input from MIDI.
- Have drums and percussion sounds assigned to MIDI channel 10 (offering at least 47 specified instruments mapped to specific keys).
- Have a minimum of 128 tones and sound effects set out according to GM categories and program change locations (see the GM tone map, page 30).
- Implement the following MIDI controllers: 1 (mod wheel), 6 and 38 (data entry), 7 (volume), 10 (pan, i.e., stereo position), 11 (expression), 64 (sustain pedal), 100 and 101 (master tuning), 121 (reset all controllers), and 123 (all notes off).
- Respond to MIDI velocity, pitch-bend, and aftertouch.
- Pitch its instruments so that Middle C corresponds to MIDI note number 60.

GM tones: instruments & program numbers. Punching up the program numbers shown on page 30 on a GM module will call up the corresponding sounds (except on MIDI channel 10, which is dedicated to playing drum sounds). If your instrument or sequencer numbers programs from 0 to 127 instead of 1 to 128, subtract 1 from each value shown.

"voices," it may only be 12-voice or even 8-voice polyphonic. Fortunately, the current trend is towards 64-voice or greater polyphony, so no matter how these instruments slice up their memory, they should be able to generate 24 *bona fide* voices. When buying an older GM module, check into how the device handles polyphony.

Though GM was a big step toward standardization, some complained because it didn't go far enough. Suppose a GM file is created while playing through one GM synth, then played back on a different GM synth. Quality can suffer because of a mismatch in balance between sounds—the violin sound on synth A might be way louder than the violin sound on synth B, for example. So if you compose a piano/violin duet on synth A and play it back on synth B, the violin part could become practically inaudible. This kind of problem shouldn't happen, but it does.

The root of the inconsistency is an incomplete definition in the General MIDI standard. When GM was defined, only the *names* of the required sounds were specified, not how they sounded. That was a necessary compromise at the time, but is now a major problem.

A limitation of GM is that 128 instrumental sounds and 47 percussion sounds aren't enough for many applications. If GM is your entire sonic world and a desired sound is not available, you have to settle for an approximation from within the GM

GENERAL PLUS

Although General MIDI provides a degree of standardization to MIDI tone modules, many people feel it doesn't go far enough. Here are highlights of two other standards that build on GM.

ROLAND GS

While Roland was careful to make GS fully compatible with the GM standard, this enhanced format—adopted so far only by Roland and several sound card manufacturers such as CrystaLake Multimedia and Orchid Technology—offers additional sounds that are selected by using MIDI program changes in conjunction with MIDI Bank Select commands.

NRPNs (MIDI Non-Registered Parameter Numbers) are harnessed to give users an element of sound-programming control over synthesis parameters such as filter cutoff frequency.

A special SFX (sound effects) set is also offered.

YAMAHA XG

The XG format expands on GS as well as GM, requiring a minimum of 32-voice polyphony, three separately controllable effects processors, plus over a dozen more MIDI-controllable synthesis parameters (such as attack time, brightness, etc.). NRPNs are also supported. While GM is limited to one bank of 128 sounds, XG defines over 100 banks of 128. There are several dedicated effects banks, which should please the gaming crowd.

---

set. Some manufacturers have tried to improve matters by creating compatible extensions to GM that provide hundreds of additional sounds. But even though the Roland GS and Yamaha XG systems try to maintain backward compatibility, a MIDI file that is optimized for these extended sets doesn't sound as good when played back on the standard GM instrument set. Furthermore, no General MIDI extension has been adopted as a standard by the industry as a whole.

No manufacturer is deliberately going to substitute a bassoon tone where the electric piano is supposed to be, but sound is subjective. Nowhere is the disparity more noticeable than on violin tones. A violin can be played sweetly and gently, or bowed fiercely. Not only will this produce vastly different sounds, but different attacks, different decays, and even different tunings. GM specifies that program 41 is a violin, and one violin is all you get.

This can be disastrous, as a violin part written for and played back on one module could sound completely different on another. Composers working in GM can either avoid such unpredictable sounds, or design sequences tailor-made for a particular GM instrument.

Another problem concerns how individual GM sounds respond to controller messages. GM proclaims that instruments should respond to a list of controllers such as

Piano

1. Acoustic grand piano
2. Bright acoustic piano
3. Electric grand piano
4. Honky-tonk piano
5. Electric piano 1
6. Electric piano 2
7. Harpsichord
8. Clavi

Chromatic Percussion

9. Celesta
10. Glockenspiel
11. Music box
12. Vibraphone
13. Marimba
14. Xylophone
15. Tubular bells
16. Dulcimer

Organ

17. Drawbar organ
18. Percussive organ
19. Rock organ
20. Church organ
21. Reed organ
22. Accordion
23. Harmonica
24. Tango accordion

Guitar

25. Acoustic nylon guitar
26. Acoustic steel string guitar
27. Electric guitar (jazz)
28. Electric guitar (clean)
29. Electric guitar (muted)
30. Overdriven guitar
31. Distortion guitar
32. Guitar harmonics

Bass

33. Acoustic bass
34. Electric bass (fingered)
35. Electric bass (picked)
36. Fretless bass
37. Slap bass 1
38. Slap bass 2
39. Synth bass 1
40. Synth bass 2

Strings

41. Violin
42. Viola
43. Cello
44. Contrabass
45. Tremolo strings
46. Pizzicato strings
47. Orchestral harp
48. Timpani

Ensemble

49. String ensemble 1
50. String ensemble 2
51. Synth strings 1
52. Synth strings 2
53. Choir aahs
54. Vocal oohs
55. Synth voice
56. Orchestra hit

Brass

57. Trumpet
58. Trombone
59. Tuba
60. Muted trumpet
61. French horn
62. Brass section
63. Synth brass 1
64. Synth brass 2

Reed

65. Soprano sax
66. Alto sax
67. Tenor sax
68. Baritone sax
69. Oboe
70. English horn
71. Bassoon
72. Clarinet

Pipe

73. Piccolo
74. Flute
75. Recorder
76. Pan flute
77. Blown bottle
78. Shakuhachi
79. Whistle
80. Ocarina

Lead Synth

81. Lead 1 (square wave)
82. Lead 2 (sawtooth wave)
83. Lead 3 (synth caliope)
84. Lead 4 (chiff)
85. Lead 5 (charang)
86. Lead 6 (voice)
87. Lead 7 (sawtooth wave in fifths)
88. Lead 8 (bass + lead)

Synth Pad

89. Pad 1 (new age, fantasia)
90. Pad 2 (warm)
91. Pad 3 (polysynth)
92. Pad 4 (space choir)
93. Pad 5 (bowed glass)
94. Pad 6 (metallic)
95. Pad 7 (halo)
96. Pad 8 (sweep)

Synth Effects

97. FX1 (ice rain)
98. FX2 (soundtrack)
99. FX3 (crystal)
100. FX4 (atmosphere)
101. FX5 (brightness)
102. FX6 (goblin)
103. FX7 (echoes)
104. FX8 (sci-fi)
105. Sitar
106. Banjo
107. Shamisen
108. Koto
109. Kalimba
110. Bagpipe
111. Fiddle
112. Shanai

Percussive

113. Tinkle bell
114. Agogo
115. Steel drum
116. Wood block
117. Taiko
118. Melodic tom
119. Synth drum
120. Reverse cymbal

Sound Effects

121. Guitar fret noise
122. Breath noise
123. Seashore
124. Bird tweet
125. Telephone
126. Helicopter
127. Applause
128. Gunshot

volume, mod wheel, aftertouch, etc., but it does not specify *how* it should respond. On one module, a mod wheel message might introduce vibrato (a slow pitch oscillation), while on another, it could introduce panning effects that cause the sound to spin crazily between the speakers. This difference in response can even occur between two sounds in the same module. And even if the mod wheel introduced vibrato on both of the two modules (or sounds), General MIDI says nothing about the vibrato's rate or depth. The results could still be very different.

For these reasons, many professionals choose Roland's Sound Canvas as a standard of comparison, and try to make sequences that sound good when played back through this module.

Standard MIDI Files. The Standard MIDI File (SMF) is a way of sharing data between different models of sequencers, even those running on different computers. Each sequencer uses its own format for saving sequence data to disk, and it would be rare for a sequencer created by one manufacturer to be able to read files created by another manufacturer's software. Most sequencers, however, will also store and load their sequences as SMFs. Some types of data (e.g., lyric or notation info) that are peculiar to an individual sequencer will be lost when the sequence is stored as an SMF, but all musical data remains intact.

SMFs are usually transferred between sequencers and between the Mac and PC platforms, or between a computer and a stand-alone sequencer, on MS-DOS-formatted 3.5" DS/DD (double-sided, double-density) disks. Many online services also offer SMFs, though concerns about copyright infringement have caused some of the larger services to eliminate SMFs from their menus.

Even stand-alone and built-in sequencers that format their disks according to their own schemes will usually read SMFs off of MS-DOS disks. (Handy hint: If you need to transfer files between two sequencers and neither of them will format disks in MS-DOS format, you can probably solve the problem by buying a box of preformatted disks at your neighborhood computer store. If they're both Macintosh sequencers, of course, this isn't necessary.)

Samplers

Instruments called *samplers* have become a vital ingredient in contemporary music-making. The concept behind sampling is simple: Plug in a microphone (or the line-level output from a mixer, CD player, VCR, DAT, etc.), record any sound, and then play it back from a MIDI keyboard. With a sampler, you can make audio collages that draw from a wide variety of sources—old movies (better get copyright clearance, though), your friends' conversation, the roar of a passing freight train—you name it.

Many computer sound cards support basic sampling, and now software is available to play individual digital recordings (samples) in response to MIDI commands. A dedicated sampler, however, is optimized for MIDI operation.

Digital sampling was first introduced as a way of triggering recordings of real sounds (violin, timpani, or dog barks) from a keyboard. In other words, samplers began as a modern electronic version of '60s tape-playback keyboards like the Mellotron. Today, they're still used for this purpose. If your sampler is equipped with a SCSI port, you can hook up a CD-ROM drive and quickly load material from a variety of discs recorded by professional sound developers. If you have a CD filled with high-quality orchestral samples, and if you understand classical orchestration, you can *almost* fool knowledgeable listeners into believing they're hearing the real thing. In fact, samplers routinely "sweeten" major movie soundtracks.

Sound developers typically sell material for samplers in both audio CD and CD-ROM format. You can record samples from an audio CD using the sampler's analog sampling input. CD-ROMs are more expensive (as are the SCSI-equipped samplers that can connect to CD-ROM drives) because they include both the audio samples themselves and program data for setting up the sampler to play the samples. Finally, some computer programs (such as Wavelab) can import samples from an audio CD loaded into the computer's CD-ROM drive. This requires a SCSI CD-ROM drive, and you still have to figure out how to get the samples into the sampler—usually via SCSI, or by saving the sample to floppy disk and loading the floppy data into the sampler.

The main advantage of a sampler over a typical sample-playback synthesizer is that the samples in a sampler play back from RAM (random access memory) rather than ROM (read-only memory). Because of this, you can load a different string orchestra for every recording session, customize the samples as required, or even plug in a microphone and make samples of noises that no synthesizer designer ever imagined. Once the sample is in the sampler, you can play it backwards, or do something like cut single words out of a spoken sentence, then paste them back together in a different order.

By sampling a beat from a record and *looping* it (that is, setting a one- or two-bar segment to play over and over), even musicians with limited technical ability can quickly put together a rhythm track. To release a song that you create by this method, you have to get the permission of whoever owns the record(s) that you sampled from (unless you enjoy being sued). If you're doing it strictly for fun, though, you can sample whatever you like.

Hip-hop and techno artists often get their beat loops from the sampling CDs mentioned earlier. These contain no complete songs, only dozens or hundreds of rhythm loops. Using such disks, you can mix and match beats to create your own song arrangements.

Like synthesizers, samplers are generally equipped with filters and envelopes for shaping the raw sampled sound. Many also include DSP algorithms (see below), with which you can process the samples in various ways. Typical DSP applications include *time-stretching*, which changes a sample's length without altering the pitch (and hopefully, without altering the timbre too much), and *normalization*, which boosts a quietly recorded sample's level so that it is as loud as possible without distorting.

Effects Processors

Electronic sound that comes straight from a synthesizer can be pretty sterile. Even an acoustic instrument track, such as guitar or conga drums, can sound naked when you listen to it just the way the microphone picked it up. The purpose of effects devices is to enhance these sounds. When effects are used subtly, the listener tends not to notice that they're being used at all; the result is merely that the sound is more natural, which helps the track "sit better" in the mix. More drastic use of effects can enrich a sound in spectacular ways, or even transform it beyond recognition.

Some synths and samplers have built-in effects processors. Free-standing processors generally come in a standard 19" *rackmount* configuration, and typically patch into a mixer's effects send and return jacks. In this setup, the effects processor's contribution to the overall sound returns to the mixer and blends with the rest of the mix. Less expensive effects are available as *stomp boxes* (floor units with built-in footswitches) that patch between a guitar and amp.

In years past, effects devices often used analog circuits or even mechanical parts, such as the metal springs in a spring reverb. These days, except for some stomp boxes, most effects are digital and use real-time DSP (digital signal processing—a fancy term for effects). DSP is not a guarantee of high quality: A cheap digital reverb still sounds cheap compared to a high-end digital reverb. But overall, the trend is toward far better sound quality and greater power for the price. The *multieffects* processor, which can produce several types of effects at once, is often a good choice for smaller studios.

The most common effect is *reverb*. Reverb simulates the echoing wash of sound in a concert hall, gymnasium, cave, or tunnel. It's important to make a distinction between reverb, which is a more or less smooth ambience, and delay, an effect that produces individual, distinct repetitions of a sound. Just to keep things confusing, both reverb and delay are sometimes called "echo."

Chorusing, flanging, and *phasing* are effects that add a rich rolling or whooshing quality to the sound. Stereo chorusing also enhances the sound spatially by spreading it across the stereo field. Although there's some overlap between the concepts, phasing (also called phase-shifting) is usually a more focused effect that is most prominent in the upper overtones of the sound, making it sound whispery or ocean-like.

Probably the most "natural" effects processor is *equalization* (EQ). An equalizer boosts or cuts selected frequencies in the audio spectrum. The usual applications are to help a track "cut through" better in a mix, reduce "boominess" in the low end, or prevent instruments in a mix from mushing together into an indistinguishable audio mess by cutting response for some instruments at frequencies that interfere with other instruments.

Equalizers designed for serious signal processing come in two main varieties— *graphic* and *parametric*. A graphic equalizer typically has a number of fixed frequency bands, each adjusted by its own front-panel cut/boost slider. A parametric usually has fewer bands of boost/cut, but offers greater control over each band.

Dynamic effects alter the loudness characteristics of a signal by restricting or expanding the available dynamic range. (Some companies use the term "dynamic effects" to refer to effects devices whose processing parameters can be controlled in real time via MIDI. While that's an important application, it's not what we're talking about here.) The most common dynamic effects, *compression* and *limiting*, can tame signals that have a few loud peaks sticking out above a generally lower level. Compressing the peaks allows boosting of the overall level, because the signal has a more even dynamic level. (Compression is often used to bring out the quieter bits of a sound, but it's important to remember that compression can alter a sound's tone color.) See Chapter 4 for more on compression.

Other categories of effects include distortion, audio enhancers, noise gating, and vocoding.

Mixers

If you have more instrument outputs that amplifier inputs, you'll need a mixer (also called a board or console; in England, it's called a desk). Mixers are indispensable in recording. They may look intimidating because of the sheer number of knobs and sliders spread across their face, but they're fairly simple to understand, because the controls for each individual channel are basically the same as those for all of the other channels—learn one channel and you've learned them all. The smallest mixers have as few as four or eight channels, while the largest have 96 channels or even more. To the right of the channel controls, a mixer will usually have a set of *master* controls, and perhaps some *submix* controls. The submix section is a separate output group that provides an additional mix to the main mix.

Let's start by looking at the channels (see Figure 19).

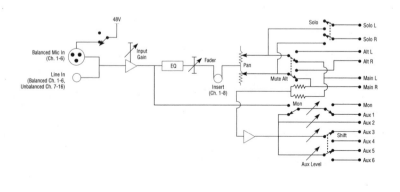

19 *This diagram shows the signal path in one channel of a 16-channel mixer. The signal enters at the left, gets processed by the EQ, faders, and whatever's plugged into the insert jack, then exits at the right.*

A channel has some sort of input jack. This links directly to an *input trim* knob, the first stage in the channel's signal path. The trim control can turn down a signal that's too loud, or boost one that's too soft, to ensure that the rest of the channel circuitry receives a signal that's at the right level for clean performance—loud enough to be above any noise, but not so loud as to cause distortion. (Sometimes the console will provide only a button to reduce the level; this is called a *pad.*) After setting the trim properly, you use the channel's main *fader* (the slider or knob conveniently located nearest the bottom edge of the mixer) to set how loud the channel's output will be in the mix.

Faders are also called *attenuators*, because they attenuate (diminish) the loudness. Setting the fader at the bottom of its throw gives *infinite* attenuation, because no sound gets through.

Above the channel fader you'll most likely find a *panpot*, which positions the channel's signal anywhere in the stereo field between the left and right speakers, and perhaps a pair of buttons for *muting* and *soloing* the channel. The mute button removes the channel from the mix without having to change the channel fader, which may already be set at the precise level called for by the song's mix. Solo allows you to hear only that channel.

Most mixers offer some EQ for each channel. The usual layout for semi-pro mixers provides low and high *shelving* EQ from a pair of knobs labeled "bass" and "treble," and one *semi-parametric* midrange band. The low and high EQ are called "shelving" because a graph of their response curve looks vaguely like a drawing of a shelf (Figure 20). With a treble shelf, the EQ boosts or cuts any portion of the signal above a factory-preset frequency (such as 10kHz). Similarly, for bass, the EQ affects any portion of the signal below a preset frequency, such as 100Hz. Some shelving EQs even let you adjust the shelf frequency.

The semi-parametric midrange band has two controls: frequency and boost/cut. A fully parametric band (Figure 21) can adjust not only the center frequency and the amount of cut or boost, but also the *width* of the affected frequency band.

Each channel will also have from two to six *auxiliary sends* (also called *effects sends*). These knobs determine how much of the channel's sound

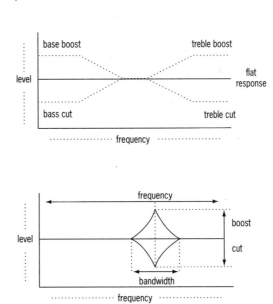

20 *Shelving equalizer response.*

21 *A parametric equalizer allows independent control over frequency, level, boost/cut, and bandwidth.*

gets sent to a separate output jack that patches to an effects processor's input. Several channels' signals can all be sent to a single effects send jack. The effects processor's output then patches into the mixer's *aux return* jacks. The master section usually includes knobs for turning the effects returns up or down. When overdubbing, aux sends can also create a separate headphone mix that the musicians can monitor. For example, a bass player might want a drums-heavy mix in order to best lock in with the beat.

A mixer designed for use with a multitrack tape deck will usually have a tape/line switch for each channel. This switch allows selecting whether the channel input comes from the line input jack or one of the tape recorder's outputs. During initial tracking (the process of recording the song), the tape/line switches will be in the line position. During overdubbing or mixdown, they will be switched to the tape position, and the faders will be used for mixing the song from the multitrack tape deck to a stereo mixdown deck.

Each channel may also have an *insert* jack. The insert is often a ¼" stereo tip-ring-sleeve jack. If no plug is in the jack, the jack is "invisible" to the channel. Inserting a plug interrupts the signal passing through the channel: the channel signal is "sent" from the tip connector, and whatever signal "returns" on the ring connector passes on through the rest of the mixer. An insert allows patching an effects processor "in-line" with a single channel without using up an effects send and return.

Once all the channels are mixed, they proceed to the *master* section. This will have, at the very least, level controls for the left and right outputs. It may also have effects return-level controls, separate output levels for headphone outs, and perhaps the output level controls for a submix.

Mixers commonly have two or more *VU meters*, which graphically show the levels for various signals ("VU" stands for "volume unit"). The VU meters may be the old-fashioned, mechanical type, with needles on a gauge, or they may have multiple LEDs. LED meters often use different colors, such as green for normal, yellow for strong signals, and red for signals that are about to clip. A single pair of meters may be switchable to show the levels at several different points in the signal path, or each channel and output bus may have its own dedicated meter.

Mixers are easy to outgrow. You'll need at least twice as many inputs as recorder tracks, and more if you use a lot of virtual MIDI tracks (i.e., MIDI instruments driven by a sequencer synced to a multitrack recorder) or do live recording with multiple mics. Note that MIDI-controllable automated mixdown accessories can help create more professional mixes, as you only have to get a mixing move right once—the automation remembers it.

Monitors

Compared to the complexities of synthesizers and computers, a speaker/amplifier system seems like a no-brainer: Turn it up, turn it down, end of story. But in fact, there are some important considerations in choosing a monitoring setup.

Using a consumer-type stereo for studio monitoring is probably a bad idea, because consumer gear is often "juiced" to sound better. Both the lows and the highs are boosted, which means you won't get an accurate picture of what's being recorded to disk or tape. On the other hand, once you've crafted a rough mix it's a good idea to play it on as many different systems as possible, including boom boxes and car stereos. If it sounds good on a bad system, it will probably sound great on a great system. For years, pro studios with $5,000 speakers have also been equipped with cheap monitor systems to simulate the sound of an AM car radio or television set.

When shopping for speakers, take your own CD, DAT, or (in a pinch) cassette to the store. Because you know exactly what the source material is supposed to sound like, it will be easier to evaluate any coloration caused by the speakers and amp.

Most of the speaker enclosures you'll run into have both a *woofer*, a large speaker to reproduce the low frequencies, and a *tweeter* to handle the highs. This is a *two-way* monitoring system; a *three-way* system has a third speaker to handle the midrange frequencies. Two-way and three-way enclosures use *crossover networks* to separate the lows from the highs, so that each speaker can efficiently do what it does best. Note that three-way systems are not inherently better than two-way systems; the value of either approach depends on the design engineer's skill, and the materials used.

For the most accurate representation of the sound, keep the speakers away from the walls and floor, since these surfaces tend to magnify the bass. Mount the speakers about one meter (three to four feet) from the position where your head will be, and about one meter apart, so that the speakers and your head form the points of an equilateral triangle. Since high frequencies in particular are somewhat directional, point the speakers directly at the listening position. Mounting the speakers closer to you rather than farther away will help minimize any coloration that may be caused by the room—often a problem in home studios.

Should you include a subwoofer in your system? That depends on the type of music you're mixing, and on the size of your main speakers. The bigger the speaker, the better the bass. If you're trying to do a dance mix through 4" speakers (or worse, headphones), you may keep cranking the bass until it's way too hot for a larger sound system. On the other hand, if you're using a subwoofer, you may be tempted to pull back on the bass frequencies, and end up with a mix that's bass-shy.

While you're picking up your new speakers, you may also want to upgrade your amp. If your amp's not powerful enough (100 watts per channel is recommended), it can be driven into distortion trying to keep up with the strong bass frequencies produced by electronic instruments. There are also monitors available with the amplification built-in (self-powered monitors). Manufacturers of these speakers claim they provide better stereo imaging as there is no crosstalk between the amp's left and right channels (as each speaker has its own amp). Also, while these speakers are heavier, you don't have to leave room for a separate amp.

It's vital that your speakers be wired in the correct polarity, with the + and − connectors hooked up the same way on both speakers. If one speaker is wired backwards, the two speakers will be *out of phase*, and the stereo image will be ruined: Sounds that you pan to the center of the field won't appear to come from the center, but from the two sides. Also, phase cancellation will cause some frequencies to disappear.

Near-Field Monitors. A near-field monitor is a loudspeaker system designed specifically to be listened to in the near field—at an ideal listening distance of about one meter. They are smaller and produce less volume than large studio monitors, and are designed to operate in imperfect acoustic surroundings (home studios).

When shopping for speakers, find a pair with a flat frequency response. Speaker frequency response is usually stated in the manufacturer's brochure, and it looks something like "20Hz–20kHz +1/−3dB." The plus-or-minus numbers tell you how much the speakers deviate from a flat frequency response. Ideally, the number should be as small as possible: +/−1dB is considered excellent.

Digital Recording: What Do I Need?

2

Know Before You Go

Setting up a digital home studio isn't something that most of us do all at once. Maybe you start out by plugging your synthesizer into the aux inputs on your stereo. Then you splurge on a MiniDisc or cassette-based multitrack recorder, and a microphone. A year later you've added a rackmount sampler and a reverb. Another six months go by and you now have a computer with a sequencer program, the stereo has been replaced by a power amp and some monitor speakers, and you're eyeing that ad for a discount DAT deck. About this time, your significant other looks at the tangle of wires and gear in the corner of the living room and says, "Isn't it about time you built a space for that stuff out in the garage?"

Looks like you're a home studio owner.

Home studios offer some wonderful advantages for musicians. You have the freedom to record what you want, when you want. You can do as many takes as you want, without running up a bill. You can learn the allied arts of recording and arranging at your own pace, without having anybody lecture you about what's right and wrong.

All the same, working in a home studio is not guaranteed to be trouble-free. Home studio owners wear a lot of hats: You have to be the engineer, arranger, and producer as well as the performer. When something goes wrong, it's up to you to fix it, and headaches caused by hardware and software problems will definitely sap your creative energy. Working alone, you may not have the benefit of feedback from anybody else, which can make the artistic, technological, and financial decisions a lot tougher. This chapter and the ones following are geared toward helping you wear your different hats with confidence. This chapter will help you decide what to put into your studio.

Develop Your Resources. Despite the wealth of information packed into this book, it can't cover everything. Check out the music sections in bookstores or libraries for more information on areas in which you're interested. Your local community college may offer courses in electronic music technology, and an internet search can yield a lot of information. Several magazines contain articles on recording, as well as reviews that can help you decide which gear is best for you.

Other musicians can be valuable resources. If you get to know musicians who have home studios, and share information with them, you'll both come out ahead. (By working together on recording projects, you can learn even more.) And it never hurts to develop a relationship with a salesperson at a nearby music store, or at several stores. By asking questions of several people, you should be able to get a well-rounded picture of the current technology. Needless to say, after taking up their valuable time, you should return the favor by making your purchases at their store rather than shopping elsewhere for the lowest price.

The Vision Thing. In planning your home studio, think seriously about your musical needs and goals. Are you looking to have a little fun? Write songs and pass them around to your friends? Is a self-produced CD the goal? Or are you aiming at a record deal and a career? Knowing how far you want to go will help you keep your equipment lust in line with reality.

And not only how far, but in what direction. What style of music will you be playing? For hip-hop and techno, a sampler with plenty of memory is probably a high priority. Try to find one that already has the maximum possible amount of memory, as an extra half-meg of sample memory can sometimes make or break a tune, and some units (especially secondhand ones) cannot easily be upgraded with currently available RAM. You may want to budget for sampler peripherals as well—a computer for sample editing, and a hard drive for file storage.

For new age recordings, high-quality reverbs may be more important, to say nothing of excellent microphones and mic preamps for recording acoustic instruments. If you plan to record experimental improvisations, a good recording deck is probably a higher priority than a powerful sequencer, and an alternate MIDI controller (or a pile of miscellaneous percussion toys) may be preferable to a master keyboard.

What will be the format of choice for what you produce? For finished pop music demos and self-produced CDs, you'll need a DAT deck or CD recorder. If you buy a rewritable (instead of write-once) CD recorder, remember that its discs are not suitable for playing back on standard CD players. Check whether the unit can record standard CD-Rs as well.

What instruments will you be recording, and what special requirements do they have? For example, if you're recording a vocalist who is less than perfect and requires multiple takes, being able to cut and paste vocals on a hard disk recording system will greatly ease the task of assembling an impressive performance. If you plan to record ensembles, how many microphones will you need? If the ensemble will be playing along with pre-recorded material, a headphone preamp with independent volume knobs is probably a must. When live musicians get into the act, the size of your room—or the number of rooms you can string cable to and from—may become an issue as well.

When budgeting for your studio, don't forget to allow more money than you think you'll need. Accessories such as cassettes, disks, connectors, cables, insurance, etc. can definitely take a bite out of your bank account.

Hardware Follies. Going out and spending your lottery winnings on 23 pieces of equipment all at once is guaranteed to produce frustration, wasted effort, and panic. Building up a studio over an extended period of time is not only financially more practical, it's also the best way to master the myriad techniques required by the various pieces of gear. MIDI is not a terribly complex technology, but when on top of MIDI you start delving into synthesizer programming, sampling, audio mixing, and composing and arranging all at the same time—oh, yeah, and also getting equipment from eight or ten manufacturers to work together smoothly—you'll have your hands full. Setting yourself up with an "instant studio" is quite likely to be a waste of money, too. After you've worked for a while with one or two synthesizers, you'll have a much clearer idea of what features are important to you and what bases your existing equipment fails to cover. When it comes to specific equipment choices, versatility is a key ingredient. Your first effects processor (software-based or not) should probably be a multieffects that can offer many different options; it shouldn't be dedicated to a single function, such as delay.

Choosing the wrong equipment is a common mistake—planning your purchases wisely isn't easy. This may sound obvious, but don't make a decision on buying a new piece until you can really audition it. If you're not sure about whether you really need something, ask if you can rent it. (Some stores will even lend gear out "on approval," so you can listen to it for a few days in your own studio and return it if its sonic performance isn't up to your standards.) Renting can also be viable if you need an item for a special project.

Getting rid of "obsolete" equipment can also be a mistake. There are times when you'll be tempted to trade in older gear for something more immediately useful, but you may find that the money gained is less important in the long run than the loss in musical power.

Thinking you always have to have the latest and greatest can be another trap. In the real world, cutting-edge technologies quite often don't work yet. The relevant variable is how much time and effort you're willing to devote to getting your system up and running.

And beware of the one-piece-does-everything system. While this can really simplify the music-making process, by definition these types of systems have limitations. Check if there are ways to interface with other pieces of gear if needed, and remember that with an all-in-one box, if it goes down you're out of business until it's fixed.

Synths and Sensibility. A single workstation-type synthesizer (keyboard, sound programs, sequencer, and built-in effects) can be a great resource, at least if you're into styles of music that involve traditional elements like melody and chords. However, any workstation—even a new one—is probably not a good choice for somebody who wants to do hardcore industrial or techno, because the sound set is not likely to be right for the next new style.

You might also want to avoid instruments that produce only one particular type of sound, no matter how popular, unless you want to specialize in playing (as opposed to recording) and plan to focus on one particular sound. Versatility is important for your introduction to synthesis.

Your first synth should be multitimbral. Buy used gear if you want to save some major money (except for certain high-priced, vintage synths). Used equipment, if it's in good condition and comes with a manual, is often an ideal place to start. If compatibility with older gear and files is an issue, don't make assumptions: Ask specific questions before you buy. In the rush to get a sexy new product out the door, manufacturers sometimes overlook the fact that their gear will be used in a studio with hardware and software that is still performing perfectly well after five years of steady use. This applies especially to waveform editing software and editor/librarians. A new synth may not be supported by any of the universal editor/librarian packages for six months to a year—or ever, if the synth isn't a success.

Used Computers. With a newer PC, a standard sound card and bundled software will give you access to both MIDI sequencing and digital audio recording. However, it's easy to forget that many musicians did just fine with a Mac Plus or even a Commodore-64. There are perfectly good MIDI sequencers that will run on an Atari ST, Amiga, Mac Plus, or a PC-XT with an early version of DOS, all of which can be picked up for next to nothing. The trick will be to find suitable software, since much of it was sold by companies that no longer exist. (Networking with other musicians will help here.) However, use old software only if it's not copy-protected. You don't want to risk losing your creative work if your only copy of the program becomes corrupted.

If you spot a used computer that was used for music or includes music programs as part of the deal, if at all possible verify that the companies that made the music software and hardware are still in business, as you may need to contact their technical support departments. Also check that any peripherals (hard disk, modem, etc.) are in good working order.

As a final thought, if your tracks don't sound good with budget gear, check out your skills as a recordist and arranger before plunking down thousands of dollars on new gear. Many hits have been cut on equipment that would now be considered laughably obsolete.

FIVE RULES FOR MUSIC EQUIPMENT SUCCESS

Rule 1: In music, the three most important factors are sound, sound, and sound. If you don't like the sound, you won't be happy with the instrument or recorder, no matter what the salesperson is trying to tell you about its features.

Rule 2: For software, the three most important factors are compatibility, technical support, and compatibility. Compatibility means not only will this software run on your computer, but it will also work with all of the other software and hardware it's supposed to work with. Be skeptical of all claims. Where possible, try installing any software and verifying that it runs before you buy it.

Rule 3: There are no dumb questions. If you don't understand how something works, ask. If the explanation you get doesn't make sense, ask more questions.

Rule 4: Shop around. Never believe a salesperson who tells you you'll get a special deal if you buy today. That's a sales tactic. Keep looking until you're sure about what you want: chances are, that same great deal will still be available next week. Of course, this doesn't apply to used gear. If it's one of a kind, when it's gone it's gone. Shop around and compare even when you're broke; that way when you do have some bucks, you'll know whether or not you're looking at a bargain.

Rule 5: Don't worry about having the latest and greatest of everything. Put your energy into making wonderful music with what you have. Sometimes the greatest musical moments come from pushing equipment past the edge because you have to.

Digital Audio Recorders

For multitrack recording, your choices have never been greater. However, you can also bet that in the future, your choices will be greater still. The audio industry evolves at a breakneck pace; today's state-of-the-art gear is tomorrow's dinosaur.

What accounts for this dizzying rate of change? Most musical products are based on digital technology—the same technology that spawned the personal computer revolution. In this highly competitive field, companies that don't come out with something newer, faster, and better every few months risk losing it all in the high-tech sweepstakes. The huge sales numbers associated with the personal computer market allow for economies of scale that keep pushing technology forward. As musicians, because our machines are based on the same "raw materials" used by computers, we benefit every time the technology moves up a notch: for example, when the price of memory decreases, we get more storage in our samplers. When hard drives decrease in price, it becomes less expensive to do digital audio (it was less than a decade ago that $2,000 was considered a great price for a 1GB, rack-mount hard drive).

As a result, in some ways this section will be out of date before the ink is dry on this page. In fact, it will be out of date before it even reaches the printer! Not to worry; this is not intended as the definitive guide to what's available, but rather, what type of specs and features you can expect for different types of gear. Besides, this information is accurate; the problem is that newer gear and updates can't be included. But you can safely assume that the gear of the future will cost less, do more, and deliver higher quality than what's available today.

So with those caveats out of the way, let's look at some real-world products circa early 1998.

Multitrack Tape Recorders

Digital multitrack tape machines broke open the budget digital recording market. The two main types are the Alesis ADAT and TASCAM DA-88. TASCAM also makes a scaled-down version of the DA-88, the DA-38, and an upscale version, the DA-98. Sony makes a DA-88 format machine, and Fostex and Panasonic have made ADAT-compatible machines. There are several different types of ADATs available at various price points.

Alesis ADAT. The basic ADAT is an eight-track recorder that digitally records audio as 16-bit data on S-VHS tape. It operates exactly like an analog tape recorder—play, rewind, fast forward, punch-in/out, etc. It also has some features you'd normally find on professional multitrack recorders, such as selectable input monitoring, auto locate, and varispeed control.

In addition to analog ins and outs, the ADAT also features a digital I/O in the form of a fiber-optic cable. This can digitally transfer eight tracks of data from one ADAT to another, as well as send data back and forth to compatible computer recording/editing systems, mixers, and synthesizers.

Up to 16 ADATs can link together to provide more tracks, with each one daisy-chained to the next via a single cable that carries sync code and transport commands. (You don't need to dedicate an audio track to SMPTE time code to synchronize multiple machines.)

The ADAT's recording medium, S-VHS tape, is cheap and reliable. An S-VHS cassette costs under $15 and offers a total recording time of up to 62 minutes, depending on what tape type (ST60-ST180) you use. Digital tape is cleaner and quieter than analog tape, and it doesn't suffer from the ill effects of generation loss when bouncing tracks or making copies.

ADAT Options. The original ADAT, which is no longer in production, was fairly basic. The unit had both −10dBV and +4dBm analog inputs and outputs, making it easy to interface in both pro and semi-pro environments. The +4dBm I/O (input/output) was via an ELCO connector; if you had a +4dBm mixing console, you needed to buy or make a special cable. Although the ADAT's native sampling rate was 48kHz, you could use variable speed to shift it down to 44.1kHz—a crude, but useable, workaround. For convenience when using multiple ADATs, the BRC (short for "Big Remote Control") served as a master synchronizer, selector, and time code generator/receiver.

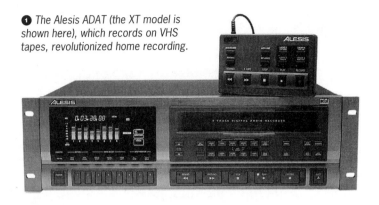

❶ The Alesis ADAT (the XT model is shown here), which records on VHS tapes, revolutionized home recording.

The ADAT-XT, also out of production (see Figure 1), updated the original ADAT with additional front-panel controls, a sturdier transport, and other extras. It was fully compatible with the original. The shuttle speed was four times faster, and several of the BRC functions were included. A 44.1/48kHz sample rate selector was added so that no manual pitch-changing was necessary. In a controversial move, all ¼" audio connectors were replaced with RCA phono jacks. The XT also had better converters: 18-bit, 128x oversampling A/D converters, and 20-bit, 8x oversampling D/A converters. (For information on buying a used ADAT, see sidebar, page 57.)

The latest ADATs have all been upgraded from 16- to 20-bit operation. The current family includes the entry-level LX20, which functionally resembles the ADAT XT but with fewer frills (for example, the analog I/O is –10 unbalanced only). The XT20 replaces the XT, and is similar with the exception of 20-bit audio. The top-of-the-line M20 is an expensive, pro-level machine designed for continuous use in recording and post-production applications. It includes a variety of bells and whistles, including extremely sophisticated sync options. The M20 is not compatible with the BRC, and has its own remote, called the CADI, that controls up to eight M20s. Studer, the famed Swiss tape recorder manufacturer, also makes a high-end machine that's very similar to the M20.

TASCAM DA-88. The TASCAM DA-88 (see Figure 2), while never selling in as great quantities as the ADAT, became a favorite of the post-production crowd and essentially took over that niche. Like the ADAT, you can lock multiple units together

❷ The Tascam DA-88 digital eight-track recorder, which uses Hi-8 videocassettes, played a significant role in the home recording revolution.

BUYING A USED ADAT

Some caveats and tips on buying a used Alesis digital multitrack recorder.

Recognizing the Variations

The earliest (oldest) ADATs are easy to recognize: the ADAT logo on the upper right is blue-black and silk-screened onto the faceplate. (The Fostex RD-8 is essentially an Original Formula ADAT with built-in SMPTE time code.) In May 1994, Alesis went to a silver plastic logo that was glued onto the faceplate. These machines delineate the transition from firmware version 3.06 to 4.0, although older machines could be upgraded to version 4.0 performance. This upgrade gave users extended features, including increased record/playback times (from 40 minutes to over 60 minutes).

To check the firmware version, hold the SET LOCATE key and press FAST FOR-WARD. The current firmware level appears on the LCD. All ADATs, no matter how old, could be upgraded to the most current firmware (version 4.03) before Alesis ceased production of its Original Formula in November 1995.

Kicking the Tires

Once you've located some likely suspects you can narrow your choices by following this advice:

Consider the overall appearance. The original ADAT chassis is constructed from sheet metal onto which the transport is mounted. Severe mechanical damage to the external case and/or a warped front panel is cause for closer scrutiny and may possibly be a disqualifying factor. (The XT uses a die-cast chassis that is extremely resistant to "warp factors.") Most ADAT top and bottom panels get scratched from being swapped in and out of racks. It ain't pretty, but it doesn't affect performance.

Check the mileage by pressing SET LOCATE and STOP. The number on the alphanumeric display indicates how many hours the tape has been in contact with the rotary heads. Thus, "0060" means sixty hours of contact, while "4234" indicates over four thousand hours. Expect to pay more for machines showing fewer hours. Be sure to ask the ADAT's previous owner about its maintenance records (check against the table on p. 84.

Who Is Selling?

There are at least three categories of ADAT user:

1. People who have been using their original ADATs heavily and are switching to newer versions for their increased wind and lock-up speeds (and in the case of the very latest ADATs, 20-bit audio converters).

2. People who need the ADAT for compatibility with the outside world, but do most of their work via MIDI and hard disk.

3. People who thought they would use the ADAT more but never got around to it.

The bottom line? An ADAT that's been on a schedule of regular preventive maintenance should chug along quite nicely. From all sellers, ask for the Operator's Manual (it's worth having), the LRC (Little Remote Control), optical and sync cables (if possible), and copies of maintenance receipts (if available).

User #1 is likely to have lots of head hours, but the machine should have up-to-date firmware and hardware. User #2 will have less time on the drum, the machine should look pretty good, be up to the latest firmware, but may need routine maintenance. User #3 will have minimal head hours, but the machine should be checked by Alesis and/or a qualified tech for firmware, circuit board revisions, the type of head, and related peripherals. A single machine, running version 3.04 firmware (for example) may behave well alone, but may not be a good team player when asked to lock up with other machines. Multiple machines should all be running the same software version for best results.

First and Second Opinion

If the seller will pop the cover, check the area around the rubber tire (between the two reel tables) for shedding. Check the pinch roller to see if it looks "glazed" and be sure the capstan is clean and shiny, not encrusted with tape oxide remnants. If necessary, use a cloth *dampened,* not saturated, with low-moisture alcohol to clean the capstan. Avoid excessive saturation because alcohol will dissolve the lubricant in the capstan bearing. For cleaning rubber parts, use Athan ATH-500-CS or a water-based cleaner, such as Windex.

Before your purchase, contact Alesis technical support for the nearest service centers in your area. Compare service charges and turnaround times. Get a serial number and confirm the unit's age with Alesis. (This may also weed out possible "hot" boxes, which you definitely want to avoid.) The primary ADAT intermittent problem is due to a dirty Mode switch. If there are no service records for the past year and a half, get on the good foot by having the switch and all rubber parts (including belts) replaced.

Eyes on the Prize

Here are a couple of basic tests to determine transport condition:

Remove the ADAT's top cover and load a tape. Once the tape is wrapped around the drum (called the Engaged mode), check basic transport functions such as play, fast forward, and rewind. Watch the tape as it moves around the guides and through the capstan and pinch roller. (You should eventually get familiar with what "normal" tape motion looks like.) Tape movement should be smooth, and there should be no slack during fast wind or spillage during stop.

For example, press Rewind in the middle of a tape and watch the take-up side. Look for slack as the machine gets up to speed. Now press Stop (check for no slack or loops), then press Stop again. The tape should disengage from the head drum assembly. Press Fast Forward (the tape is still disengaged from the head) and look for smooth travel from the supply to the take-up reel. Try these exercises at the beginning, middle, and end of the tape, looking for consistent performance at each location.

Note: If the supply and take-up reels do not come to a complete stop, the brake solenoids are either out of adjustment or have failed. If tape continues to be pulled out of the cassette shell in Stop mode, the pinch roller may not be sufficiently clearing the capstan. A minor adjustment could be all that's required, but an intermittent problem is more likely Mode-switch related.

(continued on next page)

Avoiding Moans and Groans

Loud mechanical sounds during fast forward or rewind are minor problems that can be resolved by lubricating the impedance and tachometer rollers, or by replacing the tachometer belt.

If you notice problems toward the end of a reel, suspect either supply tension, take-up tension, or pinch-roller pressure. Fast forward to about 35:00 minutes and save that as Locate 1. Fast forward to about 38:00 minutes and press Set Locate 2. Use the Auto 2 <\>> 1 and Auto Play features to continuously run this loop. The tape should be formatted, signal should have been recorded on all tracks, and you should also punch in on each track during the loop. (A portable CD player can be your source and any mixer or amp with a headphone output can be used for monitoring.)

If the error display decimal LED (located in the counter) lights up after repeated loops, either performance is marginal, damage is being done to the tape, or the tape itself is at fault. There should be no analog distortion when the machine is in Input mode, or digital noise at the punch-in/out points. If the machine passes these tests, it is in good working order as a "soloist." Next, you need to see how well it works with other ADATs.

Team Player Performance Test

This last ADAT test is for system compatibility—how the machine gets along with others. For consistent lock-ups, it is important that each machine get to the locate point at the same time, otherwise a slacker will hold up the rest of the system. When the ADAT is in unthreaded, Fast-Wind mode, the tape counter relies on information received from a reel-table tachometer. (In threaded Fast-Wind modes, it is able to accurately read time code information from tape.) Any major difference between tachometer "predictions" and tape-accurate code will cause sluggish lock-ups. To test for compatibility (this test does not apply to the Alesis XT):

1. Format two new tapes, of the same length, brand and batch.
2. Label one "Master" and the other "Slave."
3. Connect only two machines at a time using an officially sanctioned Alesis sync cable.
4. Power up the last machine first.
5. Insert tapes into the respective machines.
6. Press Locate Zero on the Master. (The slave should follow.)
7. In Unthreaded mode (press Stop twice), Fast Forward to 10 minutes and stop.
8. In Threaded/Engaged mode, each machine should be within 20 seconds of actual tape time.
9. Set Locate 1 at 10 minutes, then FF in Unthreaded mode to 40 minutes.
10. Stopped units should be within 30 seconds of actual tape time.
11. Set Locate 2 at 40 minutes, then Locate Zero in Unthreaded mode.
12. There should be 15 seconds or less difference.
13. For machines that fall within spec, make the fastest/least-hours machine the last unit in the chain.

14. The best medicine for out-of-spec machines is to schedule maintenance for "The Team." Take all the machines to one technician (at the same time) so the performance can be optimized.

Lunch: The Final Frontier

Knowing the cost of routine service is a tool you can use during the bargaining process, as well as a reality that should be factored into the cost of all tape-based systems. Perhaps before you start shopping, it wouldn't hurt to make friends with your local service facilities and…take a technician to lunch!

---

to expand the number of tracks. The machine differs from the Alesis unit primarily in that it uses Hi-8 8mm cassette tapes rather than S-VHS tapes. The two formats are thoroughly incompatible, like VHS and Beta. Think about whether you'll need to play your tapes on someone else's machine, either at a friend's house or in a major studio. If so, you'll likely want to go with the S-VHS format, as there's a large installed user base of ADATs. Hi-8 tapes are slightly more expensive than regular 8mm tape, yet they are reliable and easier to store than S-VHS tapes. One cassette can provide up to 1 hour and 48 minutes of recording time, making the DA-88 ideal for live recording.

TASCAM-family machines also have a way to transfer eight tracks of digital audio data from one machine to another, called the "TDIF" protocol. This doesn't use fiber optics, but rather, standard wire cables. Although TDIF is not as common as the ADAT Optical Interface, TDIF interface cards for digital mixers exist, and several converters allow you to convert TDIF to ADAT formats, and vice-versa.

TASCAM DA-38 and DA-98. The DA-38 also uses Hi-8 8mm recording cassettes. This lower-priced unit is specifically geared toward musicians, with an easy-to-use interface. Like the DA-88, you get 1 hour and 48 minutes recording time on a single 120 minute tape. The unit features 18-bit A/D and 20-bit D/A converters with Delta-Sigma oversampling. Options include a MIDI machine control interface (MMC-38), remote control (RC-808/848), and IF-88AE AES/EBU digital interface. The DA-98 is the highest-end digital tape machine TASCAM makes, with numerous features of interest to the post-production market, and very robust operation.

Stand-Alone Hard Disk Recorders

Disk-based systems beat digital tape for editing. You can undo, cut, copy, paste, move and sometimes even perform digital signal processing, such as EQ or time compression/expansion. The weak link for this technology is storage—once you've generated all that data, you need to save it somewhere (generally to high-capacity, removable cartridge drives or recordable CDs).

Many hard disk recorders (both stand-alone and computer-based) feature a playlist option, which allows you to string together snippets (regions) of audio data into a continuous piece of music. You can, for example, define your song's verse as one region, the chorus as another, the breakdown as another, and so on. You can then construct a song arrangement simply by entering the regions into the playlist in the desired order. You can create and edit as many playlists as desired using the same audio data, and (usually) these edits are nondestructive, resulting in no loss of the original data.

Hard disk recorders need SCSI ports to connect to external backup devices. If a SCSI port is optional, factor it into the cost—this is essential for serious work. Alternately, many machines produce data compatible with the Alesis ADAT digital tape recorder, via fiber-optic connectors, or TASCAM's TDIF protocol. You can transfer audio in either format to another machine that's equipped with a suitable interface or converter.

Digital Audio Workstations

Some hard disk recorders are built into a complete system with transport, digital mixer (sometimes with automation), equalization, and even special effects. The goal is to create a "studio in a box," and this goal succeeds to a remarkable degree. In fact, some CDs have been cut solely on these workstations. Here are some representative examples.

Roland VS-880. The latest version of the VS-880 (it's called the "V-Xpanded" VS-880; older VS-880s are software upgradable) doesn't use data compression with six-track recording, but does with eight-track. Roland's compression (Roland Digital Audio Coding, or R-DAC) is far more gentle than the ATRAC compression used by MiniDiscs. In fact, some musicians actually prefer the data-compressed sound compared to "straight" linear recording. The VS-880 has been enormously popular, and has served as an introduction to hard disk recording for many musicians (see Figure 3).

Features: Built-in 14-channel digital mixer, internal SCSI hard disk drive (can connect up to six external SCSI drives). Four analog inputs (¼" and RCA), stereo RCA auxiliary sends, RCA S/PDIF digital I/O, RCA master stereo outputs. MIDI in, out/thru. 25-pin SCSI connector. Variable gain for each input channel. 18-bit, 256X oversampling A/D converters. 18-bit, 8X oversampling D/A converters. 24-bit internal processing. 32, 44.1, and 48kHz sampling rates with variable-pitch playback. Two- or three-band EQ per input channel.

❸ The Roland VS-880, otherwise known as the "road warrior," is compact and sturdy enough to go where you need it to.

Channel muting and solo-in-place. Track-edit operations: copy, move, erase, cut, insert, exchange. MIDI control of mixer parameters. MMC, MTC, and MIDI Clock/Song Position Pointer support. Looped playback and manual, footswitch, and automated punch-in/out. Song-overwrite protection.

Capabilities: Excellent sound quality. All signals are in the digital domain. Optional VS8F-1 effects board provides 20 effect algorithms (half with three-band parametric EQ), and 200 effects patches. Three digital audio data compression modes extend the available recording time. Eight "virtual tracks" (i.e., only one can play at any time) per audio track. Non-destructive editing with 999 levels of undo. Eight locate points and 1,000 markers per song. Eight scenes (snapshots) per song, each of which retains mixer, effects, EQ, virtual tracks, and location settings. Time-compression/expansion utility time-compresses audio material down to 75% and time-expands up to 125%, with or without pitch-shift. Can choose from three time-change algorithms. Reloads the last project you were working on at startup. DAT backup of hard disk memory via S/PDIF digital connectors.

Limitations: Simultaneous recording of only four tracks. Scenes can't be sequenced to switch during a song. In linear PCM format, with no compression, the VS-880 provides only four audio playback tracks. No balanced inputs. No XLR inputs. No phantom power for condenser mics. No stereo input channels. Mixer's control complement is anemic compared to most mixers. No individual track output. The EQ adds slightly to the noise floor. If your hard disk runs out of space while you're still recording, the VS-880 can't divert the incoming audio signal to a different drive. Can't share data between songs. The LCD is small and not backlit.

Latest additions: Building on the success of the VS-880, Roland has introduced the lower-end VS-840 (very similar to the VS-880, but records to removeable cartridges and downsizes a few features) and the higher-end VS-1680, a 16-track version with a 24-bit recording mode, more inputs and digital I/O, and the ability to include up to four effects processors (as opposed to two for the VS-880 and one for the VS-840).

❹ The Fostex DMT-8VL has eight hard-disk digital tracks.

Fostex DMT-8VL

Features: Internal 1.3GB hard drive for 30 minutes of eight-track recording (can upgrade to a 2.55GB drive for 60 minutes on eight tracks). Eight channel inputs (¼" unbalanced); all channels have two-band shelving EQ. Master and monitor stereo busses, two aux sends per channel, two stereo returns, eight direct outs (RCA), RCA stereo outs, S/PDIF digital I/O (optical), MIDI in/out (see Figure 4).

Capabilities: Four-track simultaneous recording. Eight-track, 16-bit digital audio playback. DAT data backup. Cut/copy/paste editing. Auto-punch recording with rehearse. Transmits MIDI clock with tempo mapping, transmits and receives MTC. Built-in MIDI synchronization (can function as a master sync device, with keyboard and sequenced parts played "live" along with your recorded audio). Stereo digital output. Hard disk can be divided into five separate recording areas ("virtual reels"), each with its own timing information. Six memory locate points. Undo feature.

Limitations: No channel trims or XLR inputs. Only two A/D inputs. Need empty track for bouncing. No mute or solo buttons on mixer. Mixer is analog, thus has no automation. Can only access its own internal hard drive. In addition, cut/copy/paste editing in the Fostex is destructive; modified audio data has to be rewritten to the disk.

Rackmountable HD Recorders

Rackmountable hard disk recorders usually offer programmable digital mixing instead of traditional knobs and sliders. You enter commands with push buttons and view the results on an LCD.

❺ Akai's DR8 can connect to a monitor.

Akai DR8

Features: 24-bit internal processing. 18-bit 64X oversampling A/D converters, 20-bit 8X oversampling D/A converters. 32, 44.1, 44.056, and 48kHz sampling rates with variable pitch playback. Variable gain for each input channel. Nine direct-locate points, 100 auxiliary locate points. MTC sync. MMC support. Two mono effects sends, which can instead work as a stereo send. Optional video driver for onscreen graphic editing with any SVGA monitor (see Figure 5). Internal or external SCSI hard disk drives, up to seven total. Eight TRS balanced/unbalanced I/O, unbalanced stereo master outputs, XLR and RCA digital I/O connectors. 50-pin Amphenol SCSI port. Optional MIDI interface: MIDI in, out, thru. Optional SMPTE interface. Optional SCSI-B port. Each input channel has its own programmable level, pan position, send 1 level, and send 2 level settings. Three-band parametric EQ for each channel.

Capabilities: Excellent sound quality. Four-track simultaneous recording (eight with the DR16). Can sync up to eight DR8s for 64-track system. Ten edit functions: copy, copy and insert, move, move and insert, erase, delete, slip, slip track, insert, and edit undo. The DR8's internal mixer allows snapshots, crossfades between snapshots, and real-time MIDI parameter control. Manual, footswitch, and automated punch-in/out; programmable preroll. Beat/tempo map for master MIDI clock control. DAT backup of hard disk memory via AES/EBU or S/PDIF digital connectors. Backup to Alesis ADAT via optional digital interface.

Limitations: No phantom power. No XLR mic inputs.

Also available: DR16 16-track version, ADAT interface, MIDI interface, SuperView SVGA card, SMPTE reader/generator, RS-422 video sync, BiPhase film sync, SCSI-B port, MT8 Mix controller, eight-channel three-band parametric EQ, 16-channel three-band parametric EQ, and DL16 multi-machine remote control.

E-mu Darwin

Features: Internal and/or external SCSI hard drive, 44.1 and 48kHz sampling rates, eight TRS balanced/unbalanced analog outputs, four TRS

❻ E-mu's Darwin is more than just a hard-disk ADAT.

balanced/unbalanced inputs (optional expander adds four TRS balanced/unbalanced ¼" analog inputs), MIDI in/out/thru, 50-pin SCSI port, RCA S/PDIF digital I/O connectors, optional host computer SCSI interface, optional ADAT interface, optional ADAT sync card, optional DSP card.

Capabilities: Superb sound quality. Records four tracks simultaneously, allows hundreds of virtual tracks. Multiple versions per project. Non-destructive editing, up to 16 layers of undo. Forty autolocate points. Store auto locates on the fly. Stores audio data in a .WAV file compatible with Windows 95. Allows up to 2,000 edits and punch-in/out events. MIDI-controllable mixer. Transmits MMC and four MIDI Time Code formats. Optional DSP card adds 32-bit floating point processing, change-gain, fade in/out, normalize, formant pitch change, time compression/expansion. Optional ADAT sync card lets you sync up to 16 Darwins or ADATs for 16 to 128 tracks, while providing time code and word clock sync (see Figure 6).

Limitations: Editing limited to cut and paste; doesn't show waveforms. Only four analog audio inputs. No stereo line-level output. Playback stops during mode changes. Doesn't respond to MIDI continuous controllers for mixdown. No variable pitch recording or playback. No looped playback, no automated punch-in recording, doesn't slave to MIDI.

Accessories: Four-input expander, ADAT digital I/O interface, ADAT 9-pin sync interface, DSP card.

Vestax HDR-V8

Features: 18-bit A/D converters, 20-bit D/A converters, 24-bit internal bus. Ten index markers per song. Syncing of multiple HDR-V8s. Bounce up to six tracks down to one or two tracks. Eight location markers per song. MIDI clock and MTC sync. 128 mixer snapshots. Three-band EQ on each input channel. Three simultaneous aux sends per input channel. Eight balanced/unbalanced ¼" analog ins and outs, stereo

❼ *The Vestax HDR-V8 could be considered the budget-conscious HD recorder.*

¼" master outputs, RCA S/PDIF I/O connectors. MIDI in/out/thru, front panel balanced XLR mic input. Built-in 1GB hard drive.

Capabilities: The HDR-V8 can record eight tracks simultaneously and features eight virtual tracks. Built-in programmable digital mixer with snapshot recall via MIDI program changes. Can enter markers on the fly. Realtime MIDI control of mixer. Options include interfaces for the ADAT and TASCAM DA-88/38 as well as SMPTE sync, a second internal hard drive, SyJet drive for data backup, Lexicon effects, remote controller (see Figure 7).

Limitations: Song length and number of tracks fixed after first being defined. No SCSI I/O. Fixed sample rate; no varispeed recording/playback. No footswitch punch-in capability. No song copy function. Unit generates noticeable mechanical noise while in operation.

MiniDisc Recorders

These also take the studio-in-a-box approach, with an integrated mixer and MiniDisc-based multitrack recorder. They are not the preferred choice for work on major albums, jingles, or film scores due to their less than professional sound quality—all MDs use data compression to fit more data into a limited amount of storage. Typically, this compression is about 5:1, meaning that about 80% of the musical information is discarded. Remarkably, though, this influences the sound far less than you might think. Also, the data compression software (called ATRAC) is continually being refined, so while early MDs were often considered "screechy"-sounding, the latest generation is vastly smoother. Unfortunately, units cannot be upgraded easily; ATRAC is a hardware chip embedded in the MD's motherboard.

Points in the MD's favor are their lack of motor or cartridge noise, sound quality that is vastly superior to cassette-based multitracks, an extremely robust optical recording medium, portability, and ease of use. They are good for semi-pro work, multimedia music, or demos.

All of the following affordable digital multitrackers offer at least 37 minutes of four-track recording (using the 140MB MD Data Disc for four-track recording and playback, or Recordable MiniDisc for two-track recording and playback) and the ability to premix tracks without needing a free track (the MD reads the track before it writes). You can assemble playlists of songs or sections within a song, as well as set up "markers" to define specific parts of a song, then copy, delete, move, or insert these regions.

All these machines generate MIDI sync, so you can drive sequencers while you record (sync comes from within the MD and does not require giving up an audio track). Unfortunately with these recorders you cannot back up your data; you can only bounce the tracks over to another MD recorder via the analog ins and outs.

As of this writing, Yamaha has just introduced an eight-track MiniDisc. It is similar to four-track models, but offers half the available recording time.

Yamaha MD4. Four-channel mixer (mic/line), mono aux send, stereo aux return (all ¼"), stereo sub in, stereo master, and monitor outs (all RCA). Three-band EQ. Audio editing includes track/song copy, song divide and combine, cue list and song list playback. Sends MIDI clock and MTC.

❽ Yamaha's MD4 is the least expensive of the MiniDisc four-track recorders.

Capabilities: Unlike the TASCAM and Sony machines, song length is not fixed by the first track you record. Can record up to 255 songs per MD. Eight markers saved per song. Record/playback varipitch ±6%. Can record mono, stereo, and four-track songs on an MD Data disc for total times of 148, 74, and 37 minutes, respectively (see Figure 8).
Limitations: No XLR mic ins. Only one aux send. Bouncing creates digital artifacts after only a few bounces (due to an earlier version of ATRAC data compression software). Track copy can't move data to a different time location. Tempo map not saved with song. Can set monitor levels only in record mode.

Sony MDM-X4. Six-channel mixer (four mic/line plus line-only stereo channel). Two mono aux sends, two stereo aux returns, four direct track outs (all ¼" except for first two channels, which are combo XLR and ¼"). Stereo master and monitor outs (all RCA). Jog/shuttle dial. Three-band EQ on first four channels, two-band on stereo channel. Editing includes track/song copy and move, song divide and combine, undo and redo. Uses ATRAC 3.5 compression. 255 song locations per disc. Eleven mark points per song. Sends MIDI clock and MTC, receives MIDI Machine Control messages (see Figure 9).

Capabilities: Can record mono, stereo, and four track songs on an MD Data disc for total times of 148, 74, and 37 minutes, respectively. Record/playback varipitch ±8%. Can bounce three to six times without significant degradation. ATRAC version 3.5 delivers significant improvement in bounce quality over the MD4. Most powerful editing of the three recorders. Unlike Yamaha and TASCAM units can copy parts of tracks to other tracks at different time frames and among songs.

❾ Sony's MDM-X4 MiniDisc recorder includes sophisticated editing options for individual tracks.

❿ Balanced above the meter section of TAS-CAM 564 Digital Portastudio is an MD Data disc, upon which the 564 will record a total of 37 minutes of four-track audio. In some ways the 564 is like a TASCAM cassette Portastudio, but you get the advantages of editing and much improved audio quality.

Limitations: Song length is fixed by initial recording. Markers aren't saved when disc is ejected. Cue bus can't monitor mic/line input. Must set markers to see recording levels before bouncing.

TASCAM 564 Portastudio.

Eight-channel mixer (four mic/line plus two line-only stereo channels), two inserts, two mono aux sends, two stereo aux returns, cue out (all ¼"; first four channels are ¼" and XLR). Stereo sub in, stereo master and monitor outs, four direct track outs (all RCA). S/PDIF digital out (RCA). Jog/shuttle dial. Three-band EQ on first four channels, two-band on stereo channels. Editing includes song copy, index (cue list) programming, and bounce forward. Sends MIDI clock and MTC, receives MMC (see Figure 10).

Capabilities: Best mixer of the three recorders. S/PDIF digital output allows data transferral to DAT machines or computers with S/PDIF I/O for editing, EQing, or compression. Excellent jog function. Twenty index markers per song. Record/play-back varispitch ±9.9%. Can keep original or one of five takes during auto-punch recording. Instead of overwriting the data on the destination track during bouncing, TASCAM uses "bounce forward," which preserves all four original tracks (combines premixing and copying at once).

Limitations: Only 37 minutes per data disc (can't record stereo- or mono-only). Can't undo automated punch-ins. Only five songs maximum per MD (yet can access all songs on MDs formatted on the other machines). Cut/copy/paste editing affects all tracks. Song length fixed by original recording. No data sharing among songs, or between tracks of the same song. Markers aren't saved when disc is ejected.

CD Recorders

As of this writing, the latest crop of CD-R writers cost around $500 for external drives. Basically there are five drive manufacturers: JVC, Philips, Ricoh, Sony, and Yamaha. Their drives all work as CD-ROM readers as well, some at double speed (2X/2X), others at quad speed (2X/4X).

The other brands you'll see advertised are pretty much just variations on the above manufacturer's models. For example, the Pinnacle 5040 is a JVC, and the HP SureStore 4020i is a Philips. These models differ from the originals only in bundled accessories, software, documentation, and, for external units, the case, power supply, and rear-panel jacks.

CD-Rewritable (CD-RW) Drives. This new technology allows you to reuse discs up to 1,000 times. Instead of burning permanent spots onto the disc's reflective metal alloy, CD-RW technology burns semi-permanent dots, which will last (at least in theory) 30 years or more. To rewrite the data, a recording laser heats the disc, which erases the disc and makes it reflective again. Unfortunately, the dye used in CD-RW discs is only 25 percent reflective, making the disc's dots unreadable to current CD players and CD-ROM drives. However, CD-RW discs are forwardly compatible with future Multi-Read CD-ROM drives and DVD drives, whose Automatic Gain Control circuitry will compensate for CD-RW discs' low reflectivity.

A Few Audio Tips. Some recorders (including the SureStore) have provisions for copying data from a source CD to a CD-R. In such a case you won't need to pass through a sound card's digital converters. However, keep in mind that, when recording original material through your sound card's analog inputs, the CD will only sound as good as your card, so get the best that you can.

Hardware

The main differences between units are the bundled software and SCSI options.

Ricoh Rewritable 2X/6X MP6200S. The first rewritable CD recorder. Records, reads, and edits any kind of data. Can rewrite information that has already been recorded. Can burn standard CD-R discs (which are readable in most CD drives, unlike CD-RW discs) as well as CD-RW discs. Can read not only CD-R and CD-RW discs, but also video CDs, music CDs, and photo CDs. Reads data at 6X speed (900KB/sec. data transfer rate) and writes data at 2X speed. SCSI-2 interface.

Sony Spressa CDU-948S/CH. This 4X write/8X read drive for Windows 95 or NT 4.0 supports CD-DA, CD-Plus, CD-XA, CD-Bridge, Photo-CD, and Video-CD, and is CD-i ready. Writes Disk-at-Once, Track-at-Once, Session-at-Once, and Multisession, as well as packet writing. 2MB buffer memory. Includes Sony CD-Right! software. Requires Pentium processor w/ 16MB RAM, available PCI 2.1 complaint slot on board, and an open 5.25 drive. SCSI-2 interface.

HP SureStore CD-Writer Plus 7200e. Runs under Windows 95 or NT 4.0 Writes Track-at-Once, Incremental (packet), Multisession, Orange Book part 2. Bundled software includes Adaptec DirectCD, Easy CD Creator, Easy CD Audio, CD Copier, and Jewel Case Designer; Adobe PhotoDeluxe; Corel Print House Magic; DocuMagix PaperMaster Live; HP Simple Trax and Norton AntiVirus. Comes with two blank discs (one CD-RW and one CD-R). Lets you encode the folowing formats onto CD: CD-Audio (Red Book), CD-ROM (Yellow Book & Orange Book), CD-Interactive (Green Book), and Logical File Formats (ISO 9660). Minimum System Requirements: IBM or compatible with Pentium processor (75 Mhz or above), 16MB RAM.

Philips CDD3600. This 6X read/2X write recorder writes both CD-R and CD-RW discs. Supports CD-XA, Multisession, CD-i, Video-CD, CD-DA, and UDF. Writes Disk-at-Once, Track-at-Once, Multisession, and Incremental (packet). SCSI-2 or IDE interface (CDD3610). Operates on all major platforms. Windows 95 and NT compatible.

MicroNet MCDPlus4X12. This recorder (4X write/2X read) for Mac and Windows machines also comes as an internal unit (Windows only). Supports Macintosh HFS, ISO 9660, CD-ROM XA, CD-i, Video-CD, CD-Plus, Mac/ISO Hybrid, Generic, Generic XA, Mixed Mode, and CD-DA formats. Directly supports ISA bus structures, DOS, Windows, Windows 95 and Windows NT, and is Netware, OS/2, and UNIX compatible. Comes with Adaptec's Easy-CD Pro software for Windows, Toast for Macintosh.

Yamaha CDR400TXPM. The 4X/6X CDR400TXPM includes Adaptec Easy CD Pro, DirectCD, and Toast software, as well as Flash ROM software, allowing for firmware upgrades. Writes CD-DA, packet, PhotoCD, and mixed mode. 2MB data buffer minimizes possibility of buffer underruns.

JVC Personal Archiver Plus (Pinnacle 5040) and Personal ROM-maker Plus. The Personal Archiver comes in many configurations from bare-bones to full-blown, including an Adaptec SCSI card. The external unit is small and operates as a 4X reader, 2X writer. JVC's own Personal Archiver software is included for Win3.1, Win95, NT, or Mac.

Software

While CD-ROM drives are supported at the operating system level on both PCs and Macs, CD-Rs are not. Thus there's no standard software interface for CD-Rs, which means that to work well an application must contain a separate driver code for each CD-R drive it supports. Most CD-R software includes a list of compatible drives, and often recommended drives that work best.

Adaptec Toast 3.5CD-R. Toast has been the premiere CD-R pre-mastering package for the Mac since day one. It has received consistently positive reviews for its easy to use drag-and-drop interface, scripting capacities (using AppleScript), and ability to produce CD-DA format discs. Toast also creates CD-XA, CD-Video, CD-i, and 100% Blue Book compliant CD-Plus discs. Toast will work directly with both AIFF and Sound Designer II files already on disc. Supports Mac HFS, ISO 9660. It's pretty much like Adaptec's Easy-CD Pro: separate file for each track, fixed two-second preroll, test mode to avoid buffer underruns. System Requirements: 68040 or Power-Mac, System 7.5.1 or later, 8MB RAM (24MB recommended).

Adaptec Jam. This Mac CD-R software is especially geared toward creating professiona-quality audio CDs. Converts AIFF, Sound Designer II, or .WAV files into CD audio tracks. Drag-and-drop interface. Supports most CD-R and CD-RW drives. Writes in Disk-at-Once. PQ subcode editing and crossfade capability. Includes BIAS Peak LE editing software. Requires 68040 or PowerMac running System 7.5.1 or later, 8MB RAM (24MB recommended).

Digidesign MasterList CD. This program (see Figure 11) really provides what you need to burn audio CD masters reliably. The docs include a short but sweet "Successful CD Writing Guide" that includes a slew of tips and warnings, with several pages devoted to issues specific to individual drives, such as how to set the DIP switches on the Sony CDW-900E to slave multiple units on a single SCSI ID. The recommended drive list includes

⓫ Digidesign's Masterlist CD allows crossfades between tracks and a subcode editor (bottom window).

recommended media brands for each drive and also a wealth of obscure information not readily available anywhere else, like which drives don't support SCMS, or limit the number of PQ subcode changes or ISRC codes you can put on a disc. The program automatically uses disc-at-once mode when it's available, and it allows you to write at 2X or 4X speed. It has few, if any, problems with control panels or extensions. System Requirements: 68040 or Power Macintosh running System 7.5.1 or later, 8BM RAM.

⓬ *Corel CD Creator has a contemporary Windows-style interface, with a toolbar and tabbed dialog boxes.*

Adaptec Easy-CD Pro. For Win 3.1, 95, and NT, and **CD Creator** (see Figure 12) for Win 3.1, 95, NT, and Mac. Easy-CD Pro requires a hard disk with access speeds <= 19ms and sustained transfer rates >= 600KB/sec. It can create mode 2 discs (CD-XA, CD-i) and Mac HFS file formats. It also supports writing to tape, MO drives, SyQuest, and others. Another nice feature is that you can create ISO images on another system, place them on a removable hard disk, and then burn them using Easy-CD Pro. There is even a CD-i emulator. The disadvantage is that the manuals are a bit sparse; you need to already have a good idea about what you are doing if you want to create special types of CDs.

Adaptec Easy CD Creator Deluxe. This is Adaptec's flagship CD-R product for Windows 95, 98, or NT 4.0, replacing Easy-CD Pro and CD Creator for Windows products. It combines features of Easy-CD Pro and CD Creator for Windows, and includes CD Spin Doctor, an audio recording utility that lets users turn scratchy LPs into crystal-clear CDs. The new personalized compilation can play on any CD player. CD Spin Doctor can record selected tracks of music one track at a time or all at once. It supports data, audio CD, and mixed mode (data and audio) recording with a simple user interface. The package also includes Picture CD for digital photos, Video CD Creator to create video CDs, and a jewel case creator with a free CD label sampler pack. It comes bundled with PhotoSuite SE software, allowing users to create digital photo albums, photo greeting cards, personalized magazine covers, and special effect photos.

JVC's Personal Archiver Plus. (bundled with hardware; not sold separately) for Win 3.1, Win 95, NT, or Mac. Archiver supports ISO level 1 (standard CD-ROM), level 3 (up to 32 characters in a filename), Joliet (up to 64 characters in a filename—only supported by Win 95 and NT), Hi-Sierra (DOS, Unix), and Apple HFS. It doesn't support mode 2 styles such as for CD-i and CD-XA but supports single-session, CD-DA Red Book, multi-session, Disc-at-Once, Track-at-Once, enhanced music CDs (Win95), and mixed mode.

CeQuadrat WinOnCD 3.5 and Just!Audio. For Win 95 and NT (4.0 and higher) requires a minimum Pentium processor with 16MB of RAM. Supports SCSI and ATAPI/EIDE CD-R and CD-RW readers. Writes single-, multi-, and mixed sessions, Track-at-Once, Disc-at-Once; CD-XA, CD-DA, Video-CD, CD-Plus, Bootable and Hybrid CD; also DVD-ROM and DVD-video. Drag and drop interface. Just!Audio includes DeCrackler, to clean up audio from old LPs, CDs, and tapes, as well as a layout creator for artwork. Supports on-the-fly recording when using multiple CDs as source.

Media

Just as with drives, while there are lots of brands of blank CD-Rs on the market (every drive vendor has its house brand), there's only a handful of manufacturers. That short list includes Mitsui Toatsu, Ricoh, Taiyo Yuden, TDK, and Verbatim (a division of Mitsubishi); all other brands are actually made by one of those companies. Kodak is also considered a manufacturer, but it seems their discs are actually made by Verbatim—though they're not the same as either of that company's discs.

There are three basic types of discs, distinguished by the kind of dye they use. Ricoh, Taiyo Yuden, TDK, and Verbatim's regular discs use cyanine, and are often called "green" discs from the color of their bottom, dye-side layer. Kodak Info Guard and Mitsui Toatsu discs use pthalocyanine, and are often called "gold" discs. (This green-gold business can be confusing, as CD-Rs in general are also called gold discs, since the top side is always gold rather than the silver of a regular CD, and the Verbatim discs are no greener than Kodak's and Mitsui Toatsu's.) Verbatim's new Data Life Plus discs use a different technology, with a bluish metal azo dye on a silver rather than gold substrate.

If you're curious about the technology, you can find more info in the comp.publish.cdrom newsgroup and on the Web at www.cd-info.com. However, all you really need to know is that there's some controversy about the relative merits of the various types. In a nutshell, manufacturers of gold discs say they'll last longer, up to 100 years; manufacturers of green discs say they last plenty long, 70 years easy; and Verbatim says its blue discs will be dancing on the others' graves. The manufacturers and others have done accelerated aging experiments to try to support these claims, but the truth is that no one really knows how long the media will last, except that they won't last as long as regular CDs.

Which discs are compatible with your drive is a lot more important than any abstract technological comparison.

Putting It All Together. It's quite a challenge to get all this hardware and software to work in concert to produce perfect CDs every time—in fact, it's pretty much guaranteed that you'll ruin at least a few discs before you perfect the process.

Take your CD-R vendor's advice, and use only recommended discs and software—or, better yet, if possible take your CD-R software vendor's advice and buy a recommended drive. Always make sure the software is able to write in disc-at-once mode.

Recording with Computers

The desktop studio resembles the integrated studio but squeezes several functions—hard-disk-based recording and editing, customizable on-screen mixing, processing, even routing and software "plug-ins" (signal processors)—into a personal computer.

Requirements and Cautions

Processor Speed. Although you generally want the fastest possible processor speed, the computer in question may not be completely optimized for the higher processor speeds in terms of system bus speed, architecture, cache design, and speed tolerance. Anything faster than 200MHz may be running too fast for software optimized for Mac 68030 and 68040 machines. However, the faster your machine, the more tracks you can record. A 386 may only allow two tracks at a 22kHz sampling rate, while a 100MHz Pentium might play back 12 stereo tracks at 44.1kHz.

Playback Capability. Make sure you get a full-duplex board (for a PC) so you can monitor previously recorded tracks while you record new tracks.

Hard Drives. A 2GB internal hard drive is sufficient for digital audio projects, but you should really get a 3GB or more ultra-fast external drive, as large internal drives use up lots of time to search, index, run disk utilities, and/or defragment. To keep prices down, installed internal hard drives are usually slower than specialty external hard drives optimized for audio and video work. Also, keeping your audio on the same drive as your system really exercises the hard disk, as it has to run around picking up not just audio but system data. A separate drive dedicated to audio/video work is a worthwhile investment.

Not all hard drives are fast enough to record and play back digital audio. The key specs to look for, in addition to capacity, are *data throughput* and *seek time.* Data throughput relates to the speed at which you can continuously pull bytes off the disk. As long as you have a new, fast CPU, the hard drive data throughput is typically the limiting factor in determining how much data a given system can process at a time. If your computer has an older CPU, however, its speed will also limit the number of audio tracks you can play back simultaneously.

In a nutshell, if your hard disk can source a lot of data in a small amount of time, you can play back several audio tracks at once. If not, your final mix may be limited to a small number of audio tracks, and you will have to be creative to work around this limitation. Before buying a drive, check with the manufacturer of the recording software you plan to use to find out the recommended throughput. (The manufacturer will probably recommend specific drives that they have tested and know will work.)

Average seek time measures how fast (on average) a hard drive can find any given piece of data on the disk, no matter where it is reading currently. Remember, hard drives have moving parts, and we haven't conquered the limitations of physics yet. If you try to play back a number of audio files that are scattered all over the disk, the drive heads will have to jump around constantly in order to access the required data. The shorter the seek time, the less time the drive will waste on mechanical activity, and the more time it will spend transferring real data. A fast seek time is the usual distinguishing factor for an audio/video drive. Typical recommended values are around 10–12 milliseconds, but slower drives are out there, so beware. Also be careful to differentiate between average and maximum seek times. The average figure may appear impressive, but if from time to time the seek times gets much slower, you're in trouble. A maximum seek time specification lets you know the worst-case performance.

Another factor that can impact hard drive usability is *thermal recalibration.* If the drive periodically interrupts its data output to perform thermal recalibration, the audio can glitch. The best way to avoid this is to query the recording software's manufacturer for a list of recommended drives.

System RAM. The old saying that you cannot be too rich, too thin, or have too much RAM still holds true. Nothing optimizes computer usage, especially with digital audio or graphics tasks, as much as maximum system RAM. Virtual Memory and RAM Doubler help up to a point, but you really need at least 32MB of physical RAM for audio. It may be cheaper to buy less RAM with the machine and add RAM from less expensive mail-order sources after sale.

What does more RAM do for you? Here are some benefits:

- In most cases, more RAM increases your operating speed. It does this in many machines via memory access interleaving. This can mean up to a 20% speed boost on some machines.

- Having enough physical RAM to run everything you use on your computer also increases speed. Forcing a computer to go back and forth to the hard drive while performing millions of computations slows down performance. Sufficient RAM also cuts down on the use of virtual memory schemes, where the computer substitutes hard disk space for inadequate amounts of installed physical RAM.

- Sufficient RAM also allows opening up various RAM caches that speed up processor "scratch padding," operating system assist applications (such as Speed Doubler, which needs as much as 2MB), and CD-ROM pre-caching of repetitive data.

- Application size can be increased to optimize usage, and decrease the likelihood that application speed will suffer due to insufficient memory situations. If a RAM substitute such as RAM Doubler is used, it will not compensate for physical RAM needed for actual audio and video manipulations. With adequate amounts of physical RAM, however, RAM programs do work much more effectively.

- Audio and video applications will have enough room to run in RAM or in RAM disks, and to manipulate (edit) large audio and video files.

So how much RAM is enough? The minimum RAM that should be in any Pentium or Power PC computer used for recording is 32MB. More memory is better, and there are many Macs running Pro Tools or similar audio recording software that have 128 to 512MB of RAM installed. With Windows 95 machines, 64MB seems about optimum; increasing RAM doesn't produce as dramatic results as with the Mac due to limitations in the operating system. Finally, buy your RAM only from reputable local or mail order suppliers, and install it yourself only with careful and adequate static electricity protection and knowledge of your PC or Mac! You don't want to blow up your motherboard by trying to save a few pennies.

Choosing Software

MIDI-plus-digital audio. Several Mac and PC programs integrate MIDI sequencing with hard disk recording (these require reasonably capable systems with sufficient RAM and a fast hard drive). Almost all PC entry-level programs allow two tracks of digital audio and often more. For the Mac, many sequencers now take advantage of Apple's Sound Manager to record digital audio with no external hardware. Mix the sequenced sounds along with the digital audio into your final two-track master, and you have a master tape.

The software often includes mixer functions. MIDI messages can change the mix and stereo position of the sequenced tracks, as well as the digital audio tracks. So, you may need only a very simple hardware mixer to combine the MIDI sound generator outputs with the sound card's digital audio. If you use internal sounds, you may not need a mixer at all, since the sound card output will contain the final stereo output signal.

Software runs from $200 to $1,000, and an appropriate computer for around $1,000 to $2,500. A low-end sound card costs $200; a few hundred dollars more gets really good sounds and a music industry pedigree (e.g., E-mu, Ensoniq, Roland, Yamaha). However, the general cautions about hard disk recording—a big hard drive, RAM, and a backup system—still apply.

To find the right program for your studio, start surfing the web and looking for demo versions. These often do everything the regular programs do except some crucial function, such as saving data. Make sure you have at least the minimum hardware/system software necessary to run the program. (For a comprehensive comparison of digital audio recorders and their features, see *Music Technology Buyer's Guide,* from the publishers of *Keyboard.*)

Getting by with Less. Those with limited budgets can capture acoustic instruments on digital tape, bounce them over to a two-track computer-based digital audio editor for editing, then bounce back to tape for inexpensive, non-volatile storage. If you're using a small hard disk, you can then erase the disk, record more tracks, and bounce these over to tape. In a way, the tape recorder serves as a multi-purpose peripheral for the computer that provides A/D and D/A conversion as well as mass storage.

Note that MIDI tracks must all play back in real time. With tape, you can overdub the same instrument over and over on different tracks; with MIDI, if you're using instruments with lots of outputs, you need more mixer inputs.

Incidentally, MIDI isn't just for notes. A sequencer can sync to tape during mixdown to provide automated mixing and signal processing, or change presets on a guitar multieffects while you record. And if using a computer seems like overkill, many keyboard "workstations" include an onboard sequencer. This can drive not only onboard sounds but often external equipment like signal processing and automated mixdown gear. A setup consisting of a MiniDisc recorder and a keyboard "workstation" synchronized to tape can be a big price/performance winner.

A few final tips:

- Choose something that's comfortable for you. If you hate computers, a hard disk system might not be right, so digital tape would be a better choice. However, if you already have a great computer, then adding studio peripherals could be a very cost-effective option.

- Look for expandability. You'll always want more tracks, more power, more inputs and more editing features.

Choosing a Sound Card

Nearly all PCs sold today include either a pre-installed sound card, or the equivalent circuitry built into the motherboard. AV and Power Macs have built-in audio capabilities, although not a MIDI synth. Professionals usually supplement the Mac's built-in sound with additional, high-quality audio interfaces. In any event, even though virtually all sound cards advertise "CD-quality sound," in reality there's a wide range of quality.

PC sound cards are available in three basic configurations. The most common type for consumers is the *audio-plus-synthesizer* version, which contains both a MIDI synth and the ability to record and play back digital audio. Examples include the Creative Labs Sound Blaster family and the Ensoniq Soundscape. Sound card synthesizers are *multitimbral*—up to 16 different instruments can play simultaneously. (This is because MIDI supports 16 channels.) Most of today's sound card synthesizers have at least 24 voices of polyphony, but some have as many as 64 and more. More voices let you develop fuller musical arrangements and allow "layering" of sounds without having to worry about notes being cut off due to not having sufficient voices.

The second type of card, the *sampling* sound card, is generally more flexible. This lets you load in your own sounds by storing audio files (samples) in the card's RAM. It will also typically include a synth and digital audio recording/playback capabilities.

The third type of sound card, the *audio interface*, is typically a high-end product optimized for recording digital audio to the computer's hard drive. It will likely support multiple channels of audio and not include a synthesizer.

Installing Sound Cards. Connecting a sound card to a computer involves plugging the card into an expansion slot in the motherboard. Older Macs use the NuBus slot protocol, and older PCs use ISA bus-standard slots. Newer models, Mac and PC alike, use the faster PCI bus protocol that provides better throughput. However, because many ISA cards are still in use, most Windows machines offer both ISA and PCI slots.

The universality of the PCI bus has encouraged the development of *cross-platform* applications that work on both Macs and PCs. Often all that's needed to coax a piece of hardware into working on one platform or the other is a different set of *drivers*—small software routines that handle communications between the card and the computer.

Another type of slot, called PC Card or PCMCIA (Personal Computer Memory Card International Association), is usually found only on portable computers. As of this writing there are few sound cards in this format, although E-mu's new EMU8710 sound card is an exception.

When you buy a computer, the salesperson will usually know how many slots a particular model has, but be sure to ask whether they're vacant or not. Sometimes the slot count includes slots that are already occupied—by the video monitor card, for instance.

You may also encounter a board layout in which one of the ISA slots shares a chassis opening with one of the PCI slots. For example, you may get four slots of each type, but if you're using all four PCI slots, only three of the ISA slots can be used, or vice-versa.

A good sound card may need two adjacent slots, because options (such as digital I/O) often mount on a *backplate* that attaches to the computer's rear panel. Although the backplate doesn't actually plug into a slot, it nonetheless uses up the space allocated to a slot.

Unless you're planning to record to the internal hard drive, you'll probably need another slot for a SCSI adapter card. Some sound cards include on-board SCSI connectors, which can save you money as well as a slot.

IDE vs. SCSI. The original standard for PC disk drives was called IDE. Your PC's internal hard drive, floppy, and CD-ROM are probably IDE devices. IDE can be a slower data bus than the original SCSI (now called SCSI-1); the newer SCSI-2 format is faster yet. Macintosh computers have a SCSI connector on the rear panel, but PCs don't, so you'll probably want a SCSI card. SCSI-2 is backwardly compatible with SCSI-1—all you need is the right connector cable.

The most prominent SCSI card manufacturer for the PC is Adaptec, who manufacture various boards at different price points. If your software requires SCSI operation, check with the software manufacturer for a list of compatible cards before you buy one.

For digital audio applications, which can require quite a lot of data throughput, software manufacturers tend to recommend SCSI rather than IDE. To a large extent, SCSI peripherals are also compatible with both Mac and Windows systems.

IRQs & DMAs. To install hardware in a PC, you generally have to configure each card with the proper IRQ (interrupt request) number. Each hardware device in the computer requires its own unique IRQ, so that the CPU can communicate with it. If two devices share an IRQ, typically one of them will fail to work, and your computer may even lock up (however, sometimes it is possible for two devices to share an IRQ if one of them is disabled). Some cards also require DMA (direct memory access) settings.

Until recently, setting these numbers meant physically moving jumpers or DIP switches on the circuit board, as described (hopefully!) in the manual. A jumper is a tiny, horseshoe-shaped piece of plastic with a conductive metal liner. It's best handled with fingernails or needlenose pliers (and tends to fly out of your grip, especially when you're working over a shag carpet).

It's important, especially if you're still using Windows 3.1, to keep a list of which devices in your system use each of the IRQs and DMAs. If you consult the list before installing a new card, you're less likely to have to pop open the case, pull out the board, and reconfigure it.

Under Windows 95, you can use Device Manager (right-click on My Computer, select Properties, then click on the Device Manager tab) to change the IRQ settings on some boards without the muss and fuss of changing jumpers—if the board itself allows it. Device Manager will also display the currently used and unused settings for you, so keeping a list is no longer necessary (see Figure 13).

Board Handling. Even a small static discharge, such as one you can easily produce by shuffling your feet across a rug, can fry delicate electronic components.

Before touching any board, ground yourself by touching the PC's metal chassis. Some people recommend leaving the PC plugged in (but switched off) while you

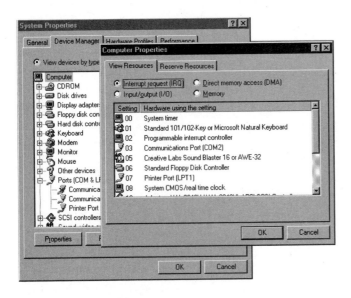

⑬ *To check your IRQ assignments, right-click on My Computer, then select Properties from the pop-up menu. This brings up the System Properties box. After clicking on the Device Manager tab and selecting Computer, click on the Properties button. This will bring up the Computer Properties box. Here, this box is being used to display the computer's current IRQ settings. Note that the Sound Blaster is set to 05, while 04 is free. To see the DMA assignments, click on the DMA radio button; Input/Output shows the memory addresses used by each device.*

install boards, as the power cable's ground connection will help protect against static discharge. If you're clumsy about dropping little screws, though, it's probably wise to unplug the computer, as you don't want one rolling around and doing something like shorting out the AC line.

After you ground yourself on the chassis, don't shuffle your feet on the rug and pick up a fresh charge. Handle the board by its backplate, or by holding edges where there aren't any components or metal traces. Don't touch the gold edge connectors along the card's bottom, and don't touch the exposed components. Insert the board with a firm, even pressure (*never* force it), and make sure that it's seated correctly before you power up again.

Also, keep the silvery envelope in which the board was packed. This is made of conductive plastic that provides protection to the board during handling and shipping. You may need to pull the board sometime—and for some weird reason, most computer stores don't stock these envelopes.

Ins and Outs. Inexpensive sound cards typically provide two line-level audio inputs (for stereo), two line-level outputs, and one or two microphone-level inputs. The line-level connectors transfer sound from cassette decks, electronic keyboards, or other standard audio devices. Since microphones generate a very low audio level, the mic input goes through a preamplifier on the sound card. Some sound cards also provide amplified speaker outputs, but you'll almost always get better audio quality by using the line-level output to feed a "real" power amp, because the onboard amplifiers are usually a cheap consumer convenience and can easily pick up the considerable electrical noise inside the computer.

Digital-Only Sound Cards. Higher-end sound cards offer digital audio inputs and outputs. These special connectors let you patch the sound card directly to compatible devices such as some CD players and DAT decks. Using these connections gives you the best possible sound, because analog/digital conversion occurs away from the noisy computer. The main limitation with digital-only cards is that this remains an analog world, so you still need to convert digital signals back into analog again before you can hear them.

One common approach is to use the converters in a DAT machine. You feed the analog signal into the DAT, put it into record mode, and patch the DAT's digital out to the sound card's digital in. During playback, you patch the sound card's digital out to the DAT's digital in, and monitor from the DAT's audio output. Unfortunately, you usually can't use both a DAT's A/D and D/A converters simultaneously; you can do A/D while the DAT is in record mode and D/A while in play mode, but not both. For stereo-only applications, like two-track editing, this isn't much of a limitation—if you need to hear what's being recorded, monitor the analog signal going into the converters.

Another approach is to use outboard A/D and D/A converters. While this is the most expensive solution, it's also the most flexible. As the quality of D/A converters improves, you can upgrade the external box without changing your sound card. Outboard converters are currently a growth industry. The quality of conversion can make a profound difference in sound quality, and many people are finding out that using outboard converters instead of the ones included in their DAT deck or sound card can result in a substantial sonic improvement. Generally, the most noticeable changes are better stereo imaging, a lower noise floor, and more "transparent" sound.

Multichannel Interfaces. Most cards internally mix all of their audio sources down to one stereo signal during recording and playback. Pricier cards support software that accommodates multiple inputs and outputs, which lets you edit and process the audio more easily. The two de facto multichannel digital audio standards are the ADAT Optical Interface (also known as the "ADAT light pipe," so called because it uses a fiber-optic cable) and TASCAM's TDIF format, which uses wires and D-connectors. These interfaces can transfer eight tracks at a time from Alesis and TASCAM eight-track digital multitrack tape recorders, respectively. You may want to look for a card that offers these interfaces as options, as you may need them later. In particular, the ADAT Optical Interface is being incorporated into more and more gear, including mixers, synthesizers, and stand-alone hard disk recorders.

Other Sound Card Considerations. Look for a card with *full-duplex* capability, which means it can play back tracks while simultaneously recording new ones.

Older sound cards used two-operator FM synthesis, which was inexpensive and sounded, well, inexpensive. (This is not to be confused with the four- and six-operator FM synthesis used in pro synthesizers.) Newer sound cards use an improved synthesis technology called *wavetable synthesis*. (If they don't have wavetable synthesis built-in, they can often be upgraded to offer it by attaching a small circuit board called a *daughterboard*.) Wavetable synthesis can produce very realistic sounds because it plays back sampled real-life instruments and sounds. The main drawback to wavetable synthesis is that the samples are typically small, as they are stored in RAM or ROM memory rather than on disk. In general, the bigger the wavetable ROM, the better-sounding the sound card. Look for a wavetable size of at least 2MB.

The *signal-to-noise ratio,* which indicates how much noise a card generates, is an important—but unfortunately, often meaningless—spec. There are many ways to measure this spec to make a card look better: for example, if a card has both analog and digital ins, the manufacturer will quote the spec for the digital I/O. Noise specs can also be unweighted (the total amount of noise) or weighted (which measures only the noise in the audible spectrum). Some companies even measure sound card noise with the card outside the computer to minimize interference. Given all this, look for the biggest number possible. 85dB is quite acceptable, especially if it's a

SOUND CARD FEATURES CHECKLIST

Here's a summary of the most desirable features to look for when shopping for a sound card. Remember to check with the sound card manufacturer to see if your computer system is compatible. And if you plan to use third-party software, be sure to ask that manufacturer which cards they recommend.

Synthesizer Specs

➤ General MIDI (GM) compatibility (additional GS or XG compatibility is better)

➤ 24-note polyphony (more is better)

➤ At least 2MB of wavetable ROM or RAM (more is usually better; RAM is also highly desirable)

➤ Effects processing (the more effects sends, the better; real-time control is useful)

➤ Programmable sounds

➤ Resonant filter

➤ Fat Seal (certifies consistent GM playback)

➤ MPU-401-compatible MIDI interface in hardware

➤ Daughterboard upgradeability

Digital Audio Specs

➤ 16-bit (or greater) resolution

➤ 44.1kHz sampling rate (additional rates are helpful)

➤ Full-duplex recording and playback (the more channels, the better)

➤ 20Hz–20kHz frequency response, with as little deviation (the "±XdB" spec) as possible

➤ 85dB (or greater) signal-to-noise ratio

Connectors

➤ The more, the better!

➤ Avoid 1/8" miniphone jacks if possible. 1/4", RCA, or (best of all) a breakout box/cable or digital interface are all better options

➤ For high-end setups, get balanced connectors. Word clock and SMPTE support, which simplify interfacing with other devices in the studio, are also recommended

"real world" spec. Well-designed cards can even push that to 95–105dB (but expect to pay more).

Regarding *frequency response,* a good sound card will cover the range from at least 20Hz to 20kHz. Even more importantly, the response curve should be *flat,* meaning that all frequencies are treated equally—there won't be any undue emphasis or loss at certain frequencies or frequency ranges. A variation of ±1dB or less is considered excellent.

Another desirable feature is realtime *signal-processing effects,* which can add an air of professionalism to any mix. Look for these on both the digital audio and the synth. 3D processing, which creates ultra-wide stereo imaging, is also becoming available.

Finally, look for cards with multiple *effects sends,* which will let you apply different amounts of processing to different tracks or synth parts. While this is a pro- rather than consumer-oriented feature, it is important for interfacing with other hardware in your studio.

Studio
Setup
and
Maintenance

3

Studio Setups

Most home studios weren't built in a day. Traditionally, home studio owners collect equipment over the years, adding new pieces and replacing
old ones according to current needs and available finances. So don't be
afraid to start small. Today's equipment is remarkably powerful; you don't need a mil-
lion-dollar control room to turn out a high-quality product, as long as you have a few
simple tools, and a healthy dose of talent. Here are four digital home studio setups,
ranging from the simple to the complex.

System One: Just the Basics

When you're just getting started, it's important to maximize the price/performance
ratio of any gear purchases. You also need to be realistic about what you'll be able to
accomplish in your studio, as well as its sonic quality.

Our first digital home studio setup is designed primarily for songwriters and composers who need to create demos of their ideas. It provides the basic tools for writing
the music, recording it, and creating a stereo cassette master. This is a good system
for people just starting out, as the financial commitment is minimal.

At the heart of the system is the music-creation tool. For keyboardists, that means
a synthesizer with a built-in sequencer. If you later find you need more voices, add a
sound module. In a basic setup, a General MIDI module would be a good choice, as
this type of unit covers lots of sonic ground. Augmenting it with a drum machine or
drum sound module would be a logical next step.

Another way to get more voices is to record keyboard parts to your multitrack
deck. For basic demos, digital MiniDisc-based multitrack recorders are excellent (if
you're on a really tight budget, cassette-based multitracks are even less expensive,
albeit less capable). Not only will they allow you to overdub multiple parts, they provide the means to add vocals, guitar, and other non-MIDI keyboard sounds to your
production. Another advantage is their built-in mixers, so you can plug your instrument directly into the same device that records the sounds. These "personal studios"

even have effects sends and returns, so you can easily route your signals through an external effects device.

One limitation of MiniDisc recorders is the ATRAC data compression they employ, and the coloration that results from *bouncing* tracks (i.e., combining the signals from several tracks into a new track, so you can record new sounds in place of the tracks you bounced). This coloration becomes particularly noticeable with multiple bounces, although the latest generation of ATRAC sounds much better than earlier versions. See "Data Storage" in Chapter 1 for more on ATRAC.

Though your synth may have built-in effects, you'll likely want to invest in an external effects device for processing vocals and other non-keyboard sounds. If you have a computer-based system, you can also use effects plug-ins to expand a program's capabilities.

Regarding vocals, unless you have an electromagnetically charged larynx, you'll need a microphone to capture your vocals on tape. Considering that this mic will be the primary conveyor of your vocal talents (as well as the main connection between acoustic instruments and your tape deck), stretch your budget a bit and go for quality. If you can afford only one mic, it should be general-purpose enough to cover duties from vocals to electric guitars. In the lower price ranges, that means choosing a *dynamic* (versus *condenser*) microphone, such as the ever-popular Shure SM58 or its near-twin, the SM57. Both can withstand high sound pressure levels (important for amplified instruments and drums), yet they're clean enough to reproduce subtleties in a lead vocal.

Unless your finances dictate buying an inexpensive, high-impedance mic, purchase a low-impedance microphone. The only concern is that most low-impedance mics have XLR connectors, which may not work with your recorder's inputs. One solution is a low-to-high impedance transformer (about $20), available at electronic shops (such as Radio Shack) and most music stores. Don't simply use an XLR-to-¼" adapter; the transformer is an essential part of the electronic conversion process. However, not all transformers sound the same, and you get what you pay for. An audiophile-quality transformer could easily run $50–$100. Fortunately, these days most mixers and recorders include at least a few balanced mic inputs, so the need to add a transformer could be moot.

Then there's the monitor system—the amp and speakers you'll be using in your studio. For this basic setup, where expense is a prime factor, consider using your home stereo (providing it's at least a step or two above a boom box). Simply plug your MiniDisc recorder's master stereo outputs into the stereo's tape or aux input, and plug your instrument(s) into the MiniDisc recorder.

The biggest problem you'll face using a home stereo is determining whether it's a reliable sonic reference point, so that music you create in your studio will sound the same when played on other systems. See Chapter 5, "Monitoring Tips," for advice on getting the most out of monitors.

One other caution when monitoring with a home stereo: Watch the volume! Keyboard instruments can produce transient signals that most home stereo speakers aren't equipped to handle. (High-energy transients are usually eliminated from CDs and tapes during the mastering process.) Unless you want to replace your speakers on a regular basis, play it safe, and monitor at reasonable levels.

The last component of your basic system is the mixdown deck, in this case a standard stereo cassette recorder. Noise reduction is important (look for Dolby C), as is the overall quality of the electronics. Avoid dual-cassette units, and those that are loaded with convenience features such as auto-reverse—more moving parts means more things that can go wrong. For best results, find a cassette deck with variable bias (this can optimize recording quality with a wide variety of tapes), Dolby HX Pro (increases high frequency response and requires no playback decoding), and three-head operation so you can monitor signals right after they've been recorded on tape, instead of hearing what's going into the deck's input (without three-head monitoring, you won't know if there are any problems until you play the tape back).

Finally, don't forget assorted cables, adapters, stands, power strips, and similar accessories. (See Figure 1.)

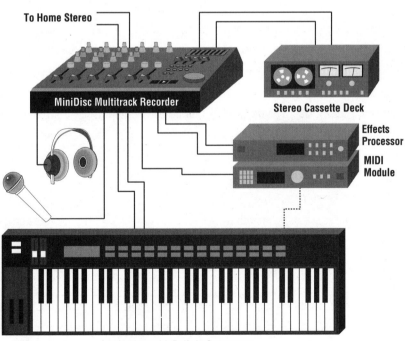

To Home Stereo

MiniDisc Multitrack Recorder

Stereo Cassette Deck

Effects Processor

MIDI Module

Synthesizer with Built-In Sequencer

❶ *System One contains everything you need to make good-sounding songwriter demos or sketch out compositions. Musical parts are created on the synth's onboard sequencer; the MIDI module provides extra voices and expanded timbral possibilities. Vocals and acoustic instruments go directly to the multitrack. There's no need for an extra mixer in this setup, as the one built into the MiniDisc recorder is more than adequate, both for recording and for final mixdown to the stereo cassette deck. Monitoring is provided courtesy of your stereo system—just keep the volume down a bit so your speakers have half a chance of surviving.*

System Two: Movin' on Up

When you need improved sonics and enhanced capabilities, you'll be faced with a dilemma—whether to add to your existing system or replace the weak links. Unfortunately, there are no easy answers; weigh each purchase on its own merits. Here are some upgrade paths to consider, and what they can do for you.

The biggest weakness in System One is the lack of quality monitoring. If you want your work to sound good on systems other than the one in your studio, return the stereo to the living room and buy a good power amp and set of near-field studio reference monitors, or powered monitors. See Chapter 1, "Monitors," for some tips on choosing the right monitor. Don't buy strictly on price or dealer recommendation. No speaker is perfectly accurate, and each has a distinctive sound. All things being equal (e.g., accurate frequency response and similar price range), choose the one that sounds best to *your* ears.

To create more complex arrangements, you'll need more synth voices and audio tracks. If your keyboard is showing its age, consider moving up to a higher-end pro instrument with more polyphony (32 or more voices), built-in effects, better keyboard "feel," and a relatively sophisticated sequencer. Alternately, try pairing a stand-alone sequencer with a less expensive keyboard. Another option is to add a MIDI sound module.

Depending on how far you want to expand your sonic palette, consider adding a sampler. Generally speaking, samplers aren't a particularly good choice when you have only one instrument, as their memory constraints usually make them less than ideal for multitimbral operation (high-priced pro models are an exception; we'll save those for a bigger system). Samplers are, however, wonderful as a secondary sound source, as they let you customize your sound in lots of ways. As an added bonus, you can record background vocals into your sampler, and easily create thick double- or triple-tracked parts simply by layering a sample with a slightly detuned or delayed copy of itself.

Even though you've added synth voices, you don't necessarily have to record them. Synth parts can be "virtual" tracks—played live by the sequencer—right up to, and including, the final mixdown. Fortunately MiniDisc-based multitrack recorders generate MIDI sync, so you can synchronize your sequenced sounds to parts (such as vocals) recorded on your MiniDisc recorder. See Chapter 4, "Synchronization," for more information.

If the MiniDisc's four tracks are too limiting, upgrade to a digital multitrack tape recorder (e.g., Alesis ADAT or TASCAM DA-88), or a hard disk recorder/workstation such as the Roland VS-880. You can expand either digital tape system to 128 tracks (16 machines running synchronously), so there's plenty of room to grow; however, these types of machines require a separate mixer. Self-contained hard disk recorders are another story, as they generally include a mixer, preamp, and even special effects.

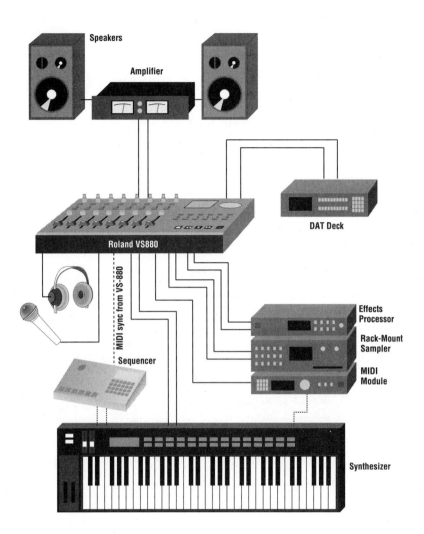

Speakers

Amplifier

Roland VS880

DAT Deck

MIDI sync from VS-880

Sequencer

Effects
Processor

Rack-Mount
Sampler

MIDI
Module

Synthesizer

❷ Upgrading from a basic setup
to a more qualified system isn't
as expensive as you might think,
provided you do things one step
at a time. Logical paths to follow:
Swap the home stereo for near-
field monitors and a good power
amp. Expand your multitrack
capabilities from four to eight
tracks. Add a rackmount sampler
for timbral expansion and sound
customization. And enhance your
MIDI editing capabilities (or at
least make your life easier) by
getting a more sophisticated,
stand-alone sequencer. Also add
a DAT deck for mixdown—you do
want to preserve your newly
found sonic quality.

Either option has tradeoffs. Digital tape's modularity makes it easy to upgrade a piece
at a time. Self-contained hard disk "ministudios" are very convenient, but what you
have is what you get (any upgrades notwithstanding).

Now is also a good time to replace your cassette mixdown deck with a DAT deck.
Consumer decks are generally less expensive than pro decks, and their audio quality is
nearly as good. You will, however, have to put up with the SCMS copy-protection
scheme found on consumer decks, which prevents you from making a digital copy of a
digital copy. This makes it difficult to do digital backups of your work, although this is
not a problem if you transfer signals via the analog inputs and outputs. (See Figure 2.)

System Three: Getting Real

Better get ready to start spending some cash, as the next leap forward involves a few
substantial investments, including a mixer, a computer, and/or a digital multitrack
recorder. We'll also look at expanding our collection of synths, mics, and effects
modules.

First up: the computer, which gives you access to powerful sequencing programs
and comprehensive editor/librarians for your synths. You'll also be able to load soft-

ware for digital audio editing, multitrack recording, and even CD recording (assuming that you have a CD recorder peripheral).

With prices plummeting in the IBM-compatible world, don't buy anything less than a fast Pentium. Equip it with a minimum of 32MB of RAM—64MB if possible—and a big, fast hard disk (at least 1GB, but preferably much larger). Windows 95/98 should be included at no extra charge.

The machine should also have plenty of full-sized slots (both ISA and PCI types) for expansion. The PCI bus is faster, but there are still lots of ISA cards around, so you really do need both types for now. Go for the biggest monitor you can afford—music software takes up substantial screen real estate, and you can never have too much viewing area. Systems are available at a wide variety of prices; shop the mail-order catalogs for some guidelines. (Consider buying locally, however, as you may get better service and faster support.) There's only one caveat: Avoid the bundled "multimedia" machines. They often come with software you'll never use, an "embedded" sound card in the motherboard that won't be as good as the sound card you'll want to add yourself, and may even use proprietary parts that could be hard to find should the computer need servicing.

In the Macintosh world, although you can easily run a number of qualified programs on older Mac models based on the 68030 or 68040 processor, the Power Mac is the way to go. Older Macs use the slower NuBus bus standard, while newer models use the PCI bus. Although PowerBooks are nifty portable computers, and as of this writing Apple is looking into creating a version optimized for digital audio, desktop machines are still best for serious audio production.

In any case, don't worry too much about CD-ROM speed, unless you're into playing games. Most of the time you'll use the CD-ROM only to install software distributed on that medium.

PC and Mac sequencing software comes in all flavors. First-timers should consider an entry-level sequencer (don't be put off by the term; most are very powerful). Most manufacturers let you upgrade from the entry-level version of their programs to the pro version for a reasonable charge, so don't feel that you'll be missing out on the good stuff if you don't dive in immediately at the upper level.

For your Mac or PC to communicate with your synths, you'll need a MIDI interface. If you want to simplify your life, several General MIDI modules have built-in MIDI interfaces, so you can take care of your polyphony and interface needs with one box.

Having a good sequencing setup isn't very useful without some MIDI modules to play back those sequences. When you move beyond Systems One and Two, you'll likely want to increase the scope of your sonic palette by looking for style- or timbre-specific sound modules.

Samplers also play an important role in expanding System Three. If you go for a used model, make sure the sampler is loaded with at least 4MB of RAM (preferably 8MB

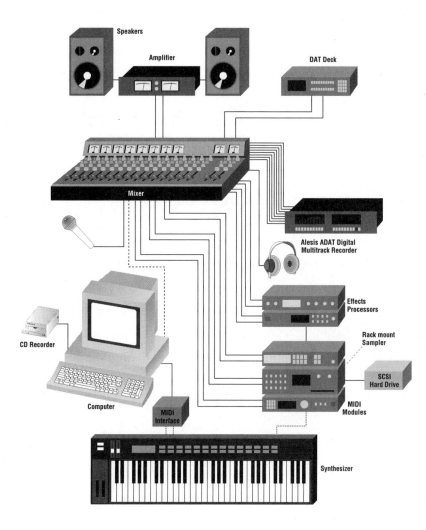

Speakers

Amplifier

DAT Deck

Mixer

Alesis ADAT Digital
Multitrack Recorder

Effects
Processors

Rack mount
Sampler

CD Recorder

SCSI
Hard Drive

Computer

MIDI
Interface

MIDI
Modules

Synthesizer

❸ Be sure you're on good terms with your banker before you start entertaining the idea of making System Three your very own. Several major purchases are involved, including a computer for sequencing and editor/lbrarian chores, sequencing/recording software, and a full-fledged mixer. You'll also need a MIDI interface. And a SCSI hard drive for sample storage. And a good microphone. If you keep adding—hard disk recording software, a digital multitrack tape recorder, compressors, exciters, microphones, and other assorted goodies— before you know it, you're the proud owner (or ower) of System Four. The digital tape recorder can record your tracks; however, if your computer is powerful enough and (preferably) has a sound card with multiple inputs and outputs, you can dispense with the multitrack tape machine and record multiple tracks directly into the computer.

or more), and that you have a large hard disk or removable media drive available for storing sample data.

For effects processing, you could move up to a programmable reverb unit. Also consider adding a compressor, which can keep dynamics under control (particularly when recording vocals, bass, and guitar).

With a suitably powerful computer, you can record directly to hard disk using a digital audio card that plugs into your computer (you won't even need a sound card for the AV or Power Macs, which have built-in digital audio). See Chapter 2, "Recording with Computers," for more on recording hardware and software.

Now is also the time to add a good condenser mic to your collection for cleaner, richer vocals, more defined acoustic guitar sounds, and a general quality boost, particularly with acoustic instruments. Most condenser mics require phantom power, so either buy a dedicated power supply, or make sure your mixer can provide phantom power.

As for mixers, you'll want at least a general-purpose (recording/PA) type with 16 input channels, built-in mic preamps, phantom power, six aux sends, four stereo aux returns, three-band EQ with sweepable midrange, and very good audio specs. Having an automation upgrade available doesn't hurt either. The next step up is a dedicated recording mixer that features tape returns and assignable subgroups.

Now that you have all this digital quality and power, a CD recorder is the next logical step. See Chapter 2 for more on CD recorders. (See Figure 3.)

Since you've upped the equipment ante, take steps to protect it. Uninterruptible power supplies, line conditioners, surge protectors, and a decent insurance policy are all in order.

System Four: Playin' in the Big Leagues

With the additions described next (all of which are considered "pro level") and enough talent, you should be able to produce CD-quality recordings, suitable for release to the rest of the world. Just hope that one of those releases is a hit—you'll need it to pay for all this gear.

First, let's look at recording media. You may want to expand a digital setup, or go for a pro-level hard disk recording system. The latter usually requires some type of audio interface (even with the Power Mac—its internal converters are no match for pro-level components), a PCI card (or NuBus or ISA card for older Macs and PCs, respectively) that links the computer to the outside and performs all of the DSP functions, and software to record and edit the audio data.

For any disk-based recording system, you need as fast and big a hard disk as possible—2GB is a minimum. Check with the manufacturer of the software you plan to use for recommended models.

One advantage of hard disk recording is that you can often use a single software program to record and edit digital audio data along with MIDI sequence data. Several companies offer these types of combined programs. While we're talking software, operators of large MIDI systems usually can benefit from a "universal" editor/librarian (and if you plan to generate lead sheets, a notation program wouldn't hurt either).

For backup, you'll need a dedicated hard disk drive for audio (removable cartridge types are a good choice), magneto-optical cartridge drive, tape drive, or recordable CD capable of recording data (not just audio).

If you have an ADAT, several companies offer interfaces so you can transfer ADAT tracks to and from hard disk recording software (there are also interfaces for the DA-88, although these tend to be less common). This means you can record on tape, edit in the computer, and mix to either tape or the computer—pretty cool.

A good high-end sampler, or synth with a sampling option, is an essential part of any pro setup. You may also find it easier to control your MIDI rig from a master keyboard. And you will need a multiport MIDI interface to drive more than 16 or 32 channels at a time, and some way to synchronize your sequencer to SMPTE time code (this is usually built into high-end interfaces).

It's probably a good idea to upgrade the effects too. For software-based systems, simply add some effects plug-ins. If your studio is more hardware-oriented, some specialty processors (tube parametric, exciter, etc.) are a good choice.

More gear also means you'll need more inputs, so think about adding a patch bay and possibly an additional *line mixer,* to submix all your keyboards to your main mixer. And while you're at it, upgrade your mic collection and monitoring system.

Maintenance
Maintaining Digital Multitrack Tape Recorders

The most important part of maintenance is preventative maintenance. Just like going in to the doctor for a flu shot before the season hits, you can prevent a lot of the problems that crop up in the studio by performing some day-to-day tasks that minimize the potential for catastrophes.

Cleaning Digital Tape Machines. Although we take them for granted, DAT machines, TASCAM DA-88's, ADATs, etc. are pretty sophisticated pieces of equipment. It takes a lot of precision to track those hair-thin tracks across a moving tape with heads rotating faster than the tires on a Subaru at 60 miles per hour. In this environment, "Cleanliness Is Next to Godliness."

Regular cleaning of your DAT (or other helical scan) machine is a must. If you wait until the "CLEAN" warning comes on or for the error LED to start flashing, then it is probably too late. The heads will have become so clogged with oxide that no amount of cleaning will repair them, and you will have to spend a lot of money to replace the head drum. A once-a-week bout with the cleaning DAT (or 8mm or VHS) tape will save you a lot of misery in the end. Please remember that the cleaning tape should only be used for about 15 seconds, and should not be rewound. By not rewinding the cleaning tape, you ensure that a new, fresh section of cleaning tape contacts the heads during the cleaning cycle. The TASCAM and Sony machines know when you have inserted a cleaning tape and go through the process automatically.

Basic Maintenance Tips
- Fast-wind all tapes end to end before formatting
- Clean the heads after formatting
- Wind tapes to either end and remove from machine when not in use
- Know how to query the machine's total head hours
- Have a maintenance schedule and stick to it, or . . .
- You know it's maintenance time if the machine eats a tape, freezes up, and displays error messages.
- Get a humidity gauge
- Don't smoke

DAT machines are pretty forgiving, but weather extremes can make your digital eight-track more temperamental than usual. Too little air moisture increases static electricity. Too much moisture makes the tape stick to the heads.

Pick up a temperature/humidity gauge and take note of the changes from day to day. Try to regulate the humidity and don't forget to change the filter in the air handler. Vacuuming is also highly recommended (be sure to change that filter, too), while smoking is not.

DA-88 Cleaning. The TASCAM/Sony digital eight-track deck has fan cooling. The fan draws air into the machine across the electronics, which is good, but it can also suck dust in through the mechanism's tape-access "port." Use the high-velocity vacuum cleaner nozzle attachment and a ½-inch artist's brush to remove the stuff that collects in this area.

ADAT Maintenance. All ADATs, regardless of brand name, are made by Alesis. Some transports are noisier than others, particularly in fast-wind modes. The two culprits in the ADAT mechanism are white plastic rollers (see Figure 4).

Noisy rollers should be replaced, but a little lubricant will shut 'em right up. Both must first be removed before lube can be applied. (Power down and unplug before removing the cover.) A white plastic cap is pressure fit over the metal shaft on which the Impedance Roller spins. Use a flat-blade screwdriver as a wedge to lift the cap. Then, *gently* squeeze the cap with a serrated-jaw long-nosed pliers, alternating clockwise and counter-clockwise while *gently* pulling (see Figure 5).

❹

❺

DIGITAL TAPE MACHINE HIDDEN FEATURES

Machine	1st Key Stroke	2nd Key Stroke	3rd Key Stroke	Function
ADAT	Set Locate & Stop			Drum "On" Time
ADAT	Set Locate & FF			Software Version
ADAT-XT	Set Locate & Stop			Drum "On" Time
DA-88	*Stop & Play			Total Drum Time
DA-88	*Stop & FF			Fast Wind Time
DA-88	*FF, Stop & Play	**Stop	Remote	Bargraph for tracks 1 & 2 will indicate errors for A & B heads when in "Play"
RC-848	While logo is displayed	Rew, FF & Stop	•First B'day	Should have v. 2.04 or higher

*The DA-88 requires the user to press these keys on Power-Up.

**Press STOP immediately after the machine is powered. The alpha-numeric display should indicate "test" mode. If so, proceed to the next keystroke.

•After installing new firmware, a "First Birthday" is required to initialize the system. Set all S1 DIP switches to On, power up, and set all S1 SIP switches to the Off position.

Let the metal shaft enter a tube of "Lubriplate" (MCM, Tel: 800-543-4330, part number 20-1325) so that a light coating is left behind. Replace the roller and gently slip on the cap until there is minimal vertical "play" in the roller. Clean the roller with a lint-free cloth dampened with 99-percent alcohol.

To access the magnetic roller, it is necessary to power the machine and coax it to lower the tape loading "elevator." This can be done via software controls.

- Press Record 1 and Record 7 while powering up.

- The front panel should display "ProG."

- Press Pitch Up until the display indicates "CAP."

- Press Auto Play to extinguish the decimal point.

- Press Pitch Down to lower the elevator.

On some transports, the tape sensing latches will keep the elevator from moving; hence take the machine to a qualified service center.

Generally Speaking. All tape machines suffer from transport-related problems. The electronics are nearly always very stable. The keys to reliable transport operation are the "Mode" and "Load" switches. These are sensors that report transport status to the system control circuitry. Dirty and worn switches generate misinformation, a.k.a. error messages and eaten or jammed tapes. If this happens to you, don't let it happen more than twice. It's time for service.

Scheduling Maintenance. Schedule maintenance every 250 hours (see table below for specific information). This is typical for the video transports used in digital tape machines. All digital eight-track decks have built-in counters that accumulate the time tape is on the heads. There are no buttons labeled "Total Head Hours." In all cases, a combination of magic buttons must be pressed to gain access to the digital netherworld.

The Panasonic SV-3700/3800 (which has a recessed, thermometer-type hour counter on the rear panel) requires the user to press "Mode, Reset, and Pause" buttons to access such need-to-know items as status (consumer or pro) and error rate. TASCAM machines must be powered up while the magic keys are pressed. (See the table "Digital Tape Machine Hidden Features," page 83, for your machine.)

ADAT MAINTENANCE SCHEDULE

Much like changing your car's oil every 3,000 miles, regular ADAT maintenance helps to avoid bigger, more costly problems later on. Even if you think your ADAT is performing normally, hold the SET LOCATE button and then press STOP to check the head-on hours.

Every 250 Head-On Hours: The ADAT's tape path and idler wheel should be professionally cleaned.

Every 500 Head-On Hours: In addition to the above, check the tape tension and pinch roller; a service center should put the ADAT through its built-in self-test routine.

Every 1,000 Head-On Hours: In addition to the above, re-align the ADAT's tape path, check the motor, and perform tests to evaluate the performance of the ADAT's digital and sync interfaces, as well as the unit's audio quality.

Every 3,000 Head-On Hours: This is the time for a transport overhaul. In addition to the above, examine the headstack and motor, and several parts should be replaced.

Warranty Coverage. TASCAM's DA-88 warranty is currently 90 days for labor and one year for parts. The Alesis warranty for ADAT XT customers is one year parts and labor; it was formerly 90 days for labor/head assembly and one year "free of defects." The XT headstack is warranted for one year or 1,500 hours. TASCAM's expected head life is 1,000 hours.

CONNECTIONS/WIRING

Cables, cords, wires—call 'em what you will, the typical home studio is filled with these long and narrow things that make no noise at all (you hope). Figuring out how everything hooks together can be pretty daunting for the newcomer.

Think of your studio's wiring not as a hopeless tangle, but as several independent networks for audio, MIDI, power, synchronization, and the computer. Here are some helpful pointers on dealing with all this.

• When rewiring or adding a new component to the system, deal with one network at a time.

• In general, power down the system before plugging and unplugging anything except MIDI cables (for audio cables, it's not necessary to power down, but you do need to turn your monitor levels all the way off). This is especially true with computer connections; you can fry a motherboard by trying to change SCSI connections (or ADB bus connections on the Mac) without shutting down.

• Keep all audio cable runs as short as possible and away from power cables. Where the two types must cross, run them at right angles, not in parallel.

• Use color-coded cables, or tape paper labels to the ends of each cable, before plugging them into the system. This will make it far easier to troubleshoot the problem when (as inevitably happens) one of your synths refuses to make a sound.

• For cable runs of more than a couple of feet, twist-ties are an easy way to group cables into snakes, and keep the rack and floor neat. Don't use the sandwich bag kind, as they have wires inside. Instead, save and reuse the plastic ones that come with your new cables. Another option: fabric stores sell Velcro® strips by the foot, and two short pieces make a convenient adjustable loop. You could even glue Velcro® patches to the back of a rack, then use Velcro® loops to corral the cables. Don't wrap snakes in tape, which can leave a sticky residue.

• Collect a box full of handy hardware accessories that make life in the studio easier. This includes plug adapters that change one plug or jack type into another, female-to-female cable extenders, and cables and/or adapters that translate male into female in all formats. Also useful are Y-cables, 9V batteries and AC adapters for guitar effects, a test tone generator, and a low-power speaker you can plug into a headphone jack (like the type intended for use with portable tape/CD players).

Even when you understand the principles of proper cabling, figuring out which cables should go where can be a challenge. Try sketching out the different kinds of recording procedures that you'll want to use in the studio. Choose your most common applications and set up your studio in that configuration. Then when you're in a creative mood, things will be set up how you want them.

Hard Disk Maintenance

Hard drives have no user-serviceable parts. However, a little preventive maintenance can help extend the life of your hard drive and possibly prevent a crash.

Preventive Maintenance

- Pay attention to early warning signs. If you hear a nasty noise when the drive is rotating, immediately back up everything! The noise probably indicates that the lubrication on the ball bearings that allow the platter to spin has broken down or dried up. Fans (replacement cost about $15) use cheaper bearings and are even more prone to this type of failure.

- Heat is the enemy. Make sure the fan/filter combination provides unobstructed air flow to prevent platter motor failure.

- To keep from losing the heads and the disk surface, avoid high impact shocks to the drive, especially while it is moving.

- Avoid having the power go off while the drive is operating (yet another argument for an uninterruptible power supply).

- Become a librarian. Since your hard drive will fail someday, always make copies of important files either on floppies, backup tapes, a second hard or MO drive, or a recordable CD. Regular backup is vital!

- Use disk diagnostic tools (such as Norton Utilities, or the Scandisk utility included in Windows 95) to test your hard drive's surface and data integrity. Do this at least once a week, or whenever an accident occurs that may affect hard drive data (e.g., the power goes out while the computer is shutting down).

- Reformat your disk periodically (but beware of copy-protected software; see below).

- Your software wants to write files to a contiguous chunk of memory. However, after repeated saves and erasures of files, it becomes harder and harder to find big chunks of memory. For example, suppose you fill up a 500MB hard drive with five 100MB files. Now you delete the second file and fourth files. Even though there are now 200MB free, if you want to save a 200MB file to disk, it will fill up the first available 100MB, then jump over to the next available 100MB and continue saving there. This creates two file fragments instead of one contiguous file. You can imagine that a drive with hundreds of files accumulated over months or years will be even more fragmented.

Fragmentation degrades disk performance because the head has to search around to find and store information. To reformat, back up all data to another storage medium, fully erase the data from the drive, reformat the drive using appropriate software (e.g., Disk Tools from the Mac's system disks), then copy back the saved data. *Caution: With Windows machines, make sure you have a way to retrieve the Windows operating system and data from the other drive.* This often involves using DOS to copy files back over. The System Disk that Windows can create under the Add/Remove Software utility will probably *not* include the drivers needed to read a CD-ROM for re-installing Windows, unless Windows came pre-installed in your machine. If you are not conversant with proper backup procedures, products such as Norton Utilities will save you much grief.

- There is still much copy-protected software for the Mac; these programs are generally "installed" on your hard drive, which sprays little invisible files around your system. *Never* defragment, optimize, or reformat a hard drive without deinstalling any copy-protected programs unless the manual specifically states that this is allowed. Otherwise, you will lose the install. Optimize only after de-installing the program(s). Use the color coding options from the Label menu to identify all copy-protected software, or when you first install the program, simply drag it over to the computer from the distribution disk or CD without officially "installing" it. This will require you to insert the program disk each time you want to run the program, but you'll be safe from installation-related problems, or losing an install if your hard drive crashes.

Newer copy protection schemes "authorize" the hard drive, and as long as the program runs on that hard drive, you're safe—even if the hard drive is defragmented. However, if you have any doubt whatsoever whether you'll lose an install by defragmenting, de-install first.

Studio AC Protection

A chain is only as good as its weakest link, and *all* your gear has one link in common: the AC supply (see Figure 6). A clean, constant supply of juice reduces stress on components, cuts noise, and can minimize mysterious power-caused glitches and crashes that sometimes plague microprocessor-controlled gear. Here are some ways to clean up your AC.

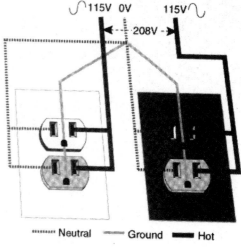

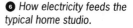

6 *How electricity feeds the typical home studio.*

- *Surge/Spike Suppressors.* A serious voltage spike (e.g., from a motor kicking in on the same line) can induce a glitch in microprocessor-based gear, and even cause physical damage. A spike or surge suppressor comes in two types: *Common-mode* and *transverse-mode.* A good suppressor offers both types of protection.

- *Isolation Transformers.* This kind of transformer isolates the gear from the AC line to provide a cleaner source of AC power. For example, MIDI Motor's Hum Buster houses the equivalent of 10 isolation transformers, with matching AC outlets, in a rack box. Because each outlet is isolated, ground loops are not possible.

- *Line Regulators.* If your power is subject to brownouts, a line regulator can help prevent losing data in RAM if there's a momentary power dip. Regulators are cheaper than uninterruptible power supplies.

- *Uninterruptible Power Supply.* A UPS contains an internal battery that is constantly being charged. If the AC input voltage goes away, an inverter processes the battery's output to provide AC power for a limited amount of time (typically 10 minutes, or more under light loads). This is usually enough time to shut down a computer system in an orderly way.

A basic industrial-grade UPS generally costs a little under $1 per watt of protection. More reserve time raises the price; consumer-oriented versions, such as those made by APS for individual computers, can go as low as 50 cents a watt.

Look for a unit with a pure sine wave output (cheaper models may generate distorted waveforms that are not suitable for sensitive gear), fast detection time (so it knows that the power has gone away), and fast transfer time (under 10 milliseconds) over to the auxiliary power. Manufacturers of these products include Sola, Stabiline, and TrippLite. Many products include insurance up to $50,000–$100,000 that covers the gear being powered, should damage occur due to electrical problems.

Where to Get AC Protection Products

Industrial-grade products are available from high-level computer stores and electronic supply houses. Consumer-oriented versions are available from mail-order computer stores such as PC Mall and MacWarehouse.

Studio Ergonomics

It's best to have all equipment needed for typical recording situations within arm's reach. Having to crawl behind or under something to access a front panel gets old pretty fast—give some thought to how you use your studio, how you flow among the various functions, and ways to minimize the amount of wasted motion that happens during the recording process.

Unfortunately, subtle ergonomic problems may go undetected, causing muscle strain, eyestrain, or minor irritation for months or years without rising above the threshold of awareness. So take a look around your studio with an eye to...

Lighting. Ideally, the main light source should be behind you to minimize glare; but position it so that your computer screen doesn't reflect the light. Lights should be bright enough that you can see what you're doing, but not so bright that the studio feels like the operating room at the veterinarian's office. (Unless, of course, you also operate on small animals as a sideline to your studio business.)

In the daytime, direct sunlight from a window can render an LED unreadable.

Likewise, a window behind the computer screen will make the screen hard to see in full daylight—but too little light on the wall behind the screen can also be a source of eyestrain. Speaking of lighting, if you need to get behind your racks to repatch modules, you may find a high-intensity flashlight a useful studio accessory.

Dimmers are definitely not recommended, unless you buy the somewhat pricier types that include built-in filtering to minimize the amount of electrical "hash" the dimmer contributes to the AC line. Otherwise, you'll probably end up with nasty buzzes when you record guitar. Sometimes dimmer buzz can even work its way into mixers and recorders. If you're in a situation where there are dimmers, turning them up to maximum brightness will cut out most of the noise.

Traditional "tube" fluorescent lights can also be a problem, but the new, compact fluorescent bulbs are a different story. These are great for the studio as they generate very little heat, seldom requirement replacement, and the light they produce is quite soft. Although expensive, the savings in electricity and replacement bulbs more than offsets the difference. They're also more environmentally friendly than traditional incandescent bulbs.

Ventilation. If you decide on an extensive remodelling job, don't forget about ventilation. Air conditioners add background rumble that's hard to remove without acoustic baffling. One tip (thanks to Spencer Brewer at Laughing Coyote Studios) is to make the ducts larger than recommended. Smaller ducts increases the amount of air pressure, whereas with larger ducts, the air just sort of "falls" out, which creates less air motion. In any event, make any contractors fully aware of your needs before they start construction.

Some studio owners swear by negative ion generators as a way to create fresher air and reduce fatigue. Regardless of whether these claims are true, these devices also remove particulate matter from the air, which is indisputably beneficial. It's generally advised not to place them too close to computers, as the negative ions could gravitate to the high voltage connections present in monitors.

Also remember that humidity isn't good for your gear (in fact, some gear, including DATs and video transport-based audio recorders, will shut themselves off automatically if the humidity goes too high). A dehumidifier is a good idea, but even "passive" dehumidifiers (such as Damp Rid) that use moisture-absorbing crystals can be tremendously helpful.

Posture. Is your chair the right height for playing the keyboard? If you're shopping for a new chair or bench, take this question into account. Not all "adjustable" stands will adjust up or down as far as they should. Depending on your setup, you may want to pay extra for an adjustable-height chair. In a larger studio, a chair with casters on the legs may make it far easier to reach all of the modules. Does the chair provide proper back support? Chronic pain in your back or shoulders after a few hours in the studio is a good sign that you may have the wrong chair. Also, the computer's mouse pad should be positioned where you can get at it without reaching or twisting your torso.

Layout. If your master keyboard has enough empty space on its top panel, you may be able to place the computer's keyboard and mouse pad there. With a two-tier keyboard stand, the computer monitor could sit on the upper tier—but this will work only if the upper tier has enough depth that the monitor can be set back, away from your face. Having the screen within 18 inches of your eyes when you're playing virtually guarantees eyestrain.

Given that the three basic components of a typical digital home studio are the computer, the master MIDI keyboard, and the mixer, one will be in front of you, one to the left, and one to the right. This puts everything within arm's reach; you can tweak on the computers or the mixer (note that with some hard disk systems, you may not need a mixer at all). Consider which element you will use most, and which needs to face the monitors in the "sweet spot," so that you can listen in stereo. If you have lots of keyboards, set up two keyboard stands off to the side as mirror images of each other, so they form a "V."

Doing control room vocals requires that you minimize ambient noise. Use external hard drives and put them below the computer table. Just putting them on the floor reduces noise; adding some carpet underneath the drives and on the underside of the table reduces the noise even more. Another noise abatement trick: use powered monitors without fans, instead of speakers and a power amp that includes a fan.

Keep guitars at right angles to the computer, which reduces noise pickup. If there's a lot of interference, just turn off the computers and go directly to tape to minimize noise (or if you use a hard disk recording system, turn off the video monitor and do your recording with keyboard equivalents).

Creative Ease. Minimize the time it takes to get your studio up and running—for example, being able to power up the whole studio from one switch is great. If you normally use two or three main pieces of software, learn how to auto-boot them when the computer switches on (e.g., in Windows, place the program, or a shortcut to the program, into the Start Up folder within the Windows directory).

Finally, keep a small, budget-priced portable cassette recorder with a built-in mic in the studio at all times, plugged in and ready to go, to "jot down" ideas quickly. Also, the small, dictation-type microcassette recorders are great for portable applications—stuff one in your shirt pocket, and you'll never lose a good idea again.

Studio Organization

A database program is a good idea for studio documentation, as it can store not only phone numbers and financial details, but what sounds are used in a given project, and where those sounds are stored. It also allows you to do keyword searches.

If you have many DATs, disks, and other storage media, a good cataloging system is essential. Label disks, and update your documentation as you go along. Later, you may not remember the details of a specific project. Many sequencer programs include a "notepad" utility, but if not, there's always the one that comes with your computer (Stickies on the Mac, Notepad for Windows). And while taking notes on your projects, if applicable, don't forget the SMPTE start time for each tune.

If you don't start each sequence track with a program change, note which program or programs are used for each track, and in which sections of the tune. If you tend to jump back and forth a lot from one tune to another, you might also find it helpful to keep an annotated, cross-referenced list of the patches currently in each synth's RAM. If you want to edit a particular patch for the tune you're working on, consult the list to find out whether that patch is being used in any other tune—and if so, the memory locations in which the new version can safely be stored. Better yet, maintain a separate sys ex disk for all the instruments used in a song.

Once you've acquired a few synths or tone modules and some extra cards or banks of sounds, organizing the sounds can become quite a challenge. Factory banks usually contain a little of everything in each bank to show off the instrument better in a store, which can make it a real chore to find what you want. After going through your available sounds and deciding which ones you'll want to keep at your fingertips, organize your RAM banks to group similar sounds together—basses in locations 1 through 8, strings in 9 through 22, and so on (or follow the General MIDI protocol if you're a fan of standardization).

However, don't fill up the entire memory (if you're using a GM-based categorization, one possibility is to leave the effects sounds empty). Keep eight or ten locations empty as storage buffers for each individual project. For example, a given tune might demand an electric piano with wild chorusing. This might be on a ROM card in a drawer somewhere, or you might have to create it by editing an existing patch. In either case, store it in internal memory at location 98 or 99. This way, you'll be able to access it with a simple program change command during the sequence—there will be no need to mess with bank select commands, or (heaven forbid) swap ROM cards in the middle of a tune.

The music isn't the only element that you may want to document; consider storing the studio configuration itself in a word processor file, detailing your equipment and how it's hooked up. This is particularly useful with multiport MIDI interfaces, so you always know which gear hooks up to which MIDI channel. Another option is to put a small, removable label on each piece of gear's front panel that shows the MIDI cable and channel number.

Recording

4

Basic Recording Tips

Many major recordings have been partially, and sometimes entirely, recorded at home studios. Granted, some of these home studios are quite elaborate, but hit records have been cut on nothing more than a beat-up Atari ST computer and a few samplers. With a little knowledge, some decent gear, a smidgen of ingenuity, and a healthy heap o' talent, you too can create great-sounding recordings at home.

Getting Started. One of the greatest advantages of having a home studio is freedom. You never know when creative inspiration may strike. Always have everything hooked up in your studio and ready to go. It's not enough just to have a studio set up; it needs to be set up efficiently, as discussed in Chapter 3.

Unless your multitrack recorder generates SMPTE time code (either by itself, or through an optional box), stripe one track with time code. Generally hard disk-based systems can generate time code; low-end digital tape machines do not, although adapters exist that can extract time code from the recorder's timing circuitry so that it's not necessary to give up a track. The time code allows synchronizing devices such as sequencers, automated mixdown modules, and even video, to your audio.

Once the preliminaries are out of the way, it's time to start tracking your instruments.

Recording Electric Guitar. Many guitar styles thrive on earthy technology, and electric guitars in particular have a limited frequency range that survives low-end recording and processing better than many other instruments. Simply put, you don't need fancy equipment to capture great guitar sounds.

Recording Direct. A recording made by miking a cool amp with a good mic in a great-sounding room tends to sound deeper and more multidimensional than any direct recording. But even with access to the above—and no volume constraints—there are times when a quirky direct sound may serve you better than a miked amp.

Guitar amps aren't neutral. A lot of what we traditionally love about electric guitar sound has to do with the idiosyncratic distortion and equalization curves they impose on your signal. Plugging your guitar directly into a mixing board or recorder input is like playing through your home stereo. You get higher highs and lower lows, but the tone may feel flat and unengaging.

If you overdrive your mixer or recorder's preamp and adjust the EQ, you start to reintroduce a guitar amp's distortion and coloration. You probably can't craft a convincing amp tone that way, but you may forge some compelling sounds. You can also create some useful direct sounds with just a few stomp boxes or a multieffects—for example, a bit of compression, some midrange dips and spikes via a graphic or parametric equalizer, and some delay can deliver a handsome clean tone.

Although the better guitar-oriented preamps and effects may not seem to have the right "feel" as you record through them, you may be pleasantly surprised by the track's authenticity during playback. Effects are especially handy for crafting an appropriate tone when adding a guitar part to an already dense track. But a less specialized preamp—just something to add sparkle and clarity to your guitar tone before it hits the mixer—may be all you need.

Miking Amps. Place the mic a few inches from the speaker. Aiming directly into the cone yields the sharpest tone; to soften it, angle the mic slightly. Another common approach is to place one mic directly on the speaker and another a few feet back. Patch both mics into your mixer, listen on headphones to the close mic alone, then bring up the distant one. You'll hear the difference instantly—the more diffuse, echoey room mic adds depth and dimension to the closer mic's sound. You also may hear a sort of hollow, phasing-like sound as you adjust the relative levels, due to possible cancellations between the two mics. That isn't necessarily a bad thing—it can add texture to the track, helping it stand out in the mix. Expensive condenser mics are

❶ *Recording from a small, overdriven amp can sound like a big amp.*

great for capturing the room sound, but you can get good results with a modest dynamic mic, or one of those flat PZM mics, which run about $50 at Radio Shack. A PZM situated on a hard floor a few feet in front of an amp can add character to a close mic, though a single close mic sometimes has the most impact.

Top-flight studios invest much time and money in tuning their rooms for optimum sound, but there's a lot to be said for exploring your workspace's random sonic idiosyncrasies. That too-full closet may be ideal for a punchy, dead sound; a small amp on a hardwood floor might offer the perfect brittleness. Try putting the amp in the shower (water off!), where the hard surfaces can lend an

eerie harshness. You can get a dirty, claustrophobic tone by miking a tiny, overdriven amp inside a cardboard box, and a garage sounds, well, garagey.

It's tough to beat the tone of a modest combo amp. Obviously, you can get power-amp distortion at lower volume with a small amp. But don't overlook the possibility of running a distortion pedal into a quiet, clean amp.

Don't rule out using just plain bad amps, which can lend a unique edge and character. Try playing through a cassette player, an intercom, or a walkie-talkie. And those toy Marshall half-stack amps are superb for creating hard, pinpoint tones that can slice through a dense mix (see Figure 1).

Recording Acoustic Guitar. To capture the true sound of a fine acoustic, nothing beats an expensive condenser mic. If you specialize in solo acoustic playing, your best bet is probably to find a great-sounding room—such as a church, or a woody-sounding front parlor—and record direct to DAT through a pair of condenser mics and a couple of quality preamps (many guitarists favor tube preamps for their "warmth").

Direct recording options are less appealing than for electric. The piezo-electric pickups built into many new acoustic-electrics are controversial. Some guitarists revel in their crisp sound; others find even the best systems irritating and quacky. Usually, proper equalization can bring up the lower midrange and bass, improving the sound dramatically. However, even unequalized piezo pickups can sound just fine when laid back in a mix, and that characteristic piezo "zing" can be just what's needed to stand out in a track.

Aiming a mic directly at the soundhole usually sounds too boomy. Try angling it toward the neck-body junction—even then you may have to roll off some low midrange frequencies. And because the acoustic's dynamic range is so much greater than an overdriven electric guitar, you may need some compression to capture soft passages at an adequate level without having loud parts exceed the available dynamic range.

Sticking a magnetic soundhole pickup on an acoustic and playing through an amp or amp simulator is a fascinating and under-explored approach. It doesn't sound anything like a miked acoustic guitar, but it can yield some spectacular sounds.

Using Effects with Guitar. Global warning: Effects tend to obscure the character of your instrument and the way you play it. Don't be afraid of them, but be aware of their tendency to make your guitar sound generic. Guitar tracks often need less distortion than you might think. Excessive distortion tends to wash out the guitar's entire frequency range. Context is everything—a moderately distorted tone can seem like the filthiest thing in the world after a comparatively clean passage.

Modulation effects like chorus, flange, and tremolo can animate a track or just dull your point with unnecessary detail. Chorusing thickens, but generally removes edge, though the animation of chorusing and flanging might draw attention to an instrument obscured in a mix. If you use a flanger you should probably record it to the multitrack rather than add it during the mix. This way you can record the sweep you like best. Another option is to sweep the effect manually while mixing, although this may be a problem if you're occupied with other aspects of the mixing process.

Think dynamically. For example, instead of using a bit of reverb throughout, try a relatively dry mix with just one intermittent part subjected to heavy reverb—that way you'll at least have some spatial and temporal contrasts. And don't be afraid to explore your own spaces—thousands own the same reverb chip, but only you have your tiled bathroom.

If you think you need reverb, first consider the delay alternative. Delays can add the spaciousness you seek without gumming up the entire track. Old-style analog delays have diminished high-end response, which makes them sound warmer, with the fuzzy echoes sitting neatly behind the direct guitar signal. You can fake that sound by splitting your guitar into two signal paths. One goes directly to the mixer, while the other goes to the mixer through the delay (set this for delayed sound only—no straight sound). Roll off some of the highs on the mixer channel carrying the delayed signal; you might also want to experiment with stereo placement (e.g., guitar panned left of center, with the echo panned right of center). Also try recording simultaneous miked and direct tracks, and separate them slightly in the mix.

If you have tracks to spare and want a "wider" sound, try doubling a part. Panning an exactly doubled part full left and right—a technique favored by everyone from acoustic strummers to death-riffers—is guaranteed to create an ultra-wide stereo image, but that extreme approach can get fatiguing, especially on a loud foreground part. Consider narrower panning, and don't rule out doubling in mono. But keep in mind that doubling doesn't necessarily make a part sound *bigger*, just wider.

Finally, don't forget to inflict your funky guitar effects on other instruments. Controlled distortion, for example, can sound amazing on drums, or transform a weenie keyboard preset into a terrifying roar. And you may be shocked by how wicked vocals sound through a dedicated guitar reverb—or a guitar amp.

The digital tape/hard disk explosion has spurred a renaissance in tube preamps and compressors designed to inject analog warmth, with a number of new devices aimed at the semi-pro market. These boxes can soften the sometimes clinical contours of digital recordings. You can record individual tracks through them—perhaps going straight to the recorder, bypassing the mixing board—or use them to warm up an entire mix. They impart a more tape-like sound to the individual tracks, as well as help them merge better.

Great mixes are dynamic. A nice sound isn't enough—it has to move. Give your mix a foreground and background; not everything can be loud. Explore contrasts of wet and dry, as opposed to cramming everything into one space, however attractive. Don't be afraid to use small sounds, which give big sounds room to exist (and make them sound huge in comparison). If the guitars are big, try small drums, and vice versa. Play left against right, near against distant, loud against soft, bright against warm, smooth against ugly.

Don't be a perfectionist about mistakes. So many guitarists let their inspiration dribble away while chasing perfect takes. We tend to be more forgiving of others' flaws than our own—are you really bummed out by the small clams you hear on so many classic records?

Finally, don't be afraid of your naked tone. Let your fingers and frets come through; they are more compelling than any mass-produced gadget. Your personality is your greatest asset.

Recording Vocals. Vocals are another problem area for most home studios. Once again, it's hard for the humble home studio to compete with the expensive microphones, preamps, and outboard gear used by most professional studios. But with a modest investment in the proper equipment and a little experimentation, it is possible to achieve good-sounding vocals at home. Get the best mic you can afford, and use (but not abuse) a compressor/limiter.

Try and use separate mics for each vocalist, so you have more control over the mix. If your recorder has a VSO (variable speed oscillator, sometimes called a pitch or speed control), you can use this to create chorus, harmonizing, or equalization effects that sound particularly effective on background vocals. For example, for a fuller sound, record the first vocal track at normal speed, the second track about 5% slower, and the third track 5% faster.

Experiment with different microphone positions before going for a killer take. Microphones can sound vastly different depending upon where they're placed in relation to the sound source, as well as reflective surfaces in the room. Once you've found the right spot, mark it with masking tape on the floor. It's also usually not best to record a vocalist by putting the mic right in his or her face. It should be at least eight or ten inches away, and a bit above the mouth in order to avoid pops and sibilance. A foam-rubber pop screen, available at any Radio Shack, is a must.

Recording Drums. It is very difficult to record acoustic drums in a home studio. Even if you're lucky enough to have a large, good-sounding room to record in, you'll also need thousands of dollars' worth of microphones, preamps, and processors to achieve optimum sound quality—not to mention neighbors who can handle the noise.

Because of this, most people use drum machines in their home studios, but even these can be problematic. The biggest complaint is that drum machines often sound too much like machines. However, there are several strategies you can use to make your drum parts sound more realistic. See "Recording Electronic Drums" below, for tips on improving the sound of your drums.

About Effects. The source should sound as good as possible before you record it. Use effects sparingly; when you sit in your studio for hours, after a while you'll get used to the reverb, so you'll want to add more. Eventually, you'll be swimming in reverb. Over-equalizing changes phase relationships to other sounds. You might want to use EQ to cut rather than boost; for example, if a sound isn't bassy enough, consider rolling off the treble rather than boosting the bass. This will give the EQ more headroom, and reduce the chance of distortion.

If you want a track with more presence and headroom, try using a compressor (but don't overcompress, or the sound will be "smaller"). A noise gate, designed to eliminate tape hiss during moments when a given track is silent, can add space by removing unwanted noise or shortening decays.

Experiment with your signal processors. Although a pitch shifter is designed to create a clone of a track at an altered pitch, instead try using it to move the vocalist's reverb or delay a fifth above the pitch at which he or she is singing. Or, use a noise gate to eliminate a keyboard pad during the spaces between hi-hat hits (using the gate's key input).

Many engineers compress the full mix as a matter of course, particularly with beat-heavy music, to allow the kick and snare to dominate the sound without pushing the signal into overload. This is less effective with "bands" made up of sampled and synthesized sounds, because they often start out tremendously compressed. And, of course, it may not be musically appropriate in other situations.

Try recording tracks with effects, and make sure the performer can hear the effects while playing. In these days of multi-processors and affordable multitracks, this isn't common practice; it's more likely that decisions about which effects to add will get postponed until mixdown time. But effects—delay, flanging, and reverbs of various sorts—can influence the performance, and you may well find that the two are better married if they're performed and recorded together. Also, a board that lacks lots of effects sends and returns for each channel will limit how many effects you can add during mixdown.

Performance vs. Perfection. All the emphasis on perfection nowadays has overshadowed one of the most important aspects of music—performance. While it doesn't hurt to spend a little extra time making sure everything sounds good and that you don't make any mistakes, always keep focused on the emotion and energy you're trying to convey.

Final Tips. When recording digitally, watch your levels. Once you hit true zero, it's all over (load). On the other hand, don't be too conservative or you'll lose valuable resolution. Every 6dB down from zero is the equivalent of giving up one bit of resolution. Therefore, if your peaks don't exceed –6 with a 16-bit recording medium, then your recording is actually a 15-bit recording, not 16-bit. However, if a piece of digital audio has just a few extremely short transients that hit maximum headroom, you can go a little over zero and (hopefully) not hear the difference.

Practice routine studio maintenance. With analog tape, clean those tape heads before every session. With digital tape, clean the heads with a dry cleaning cassette every 100–200 hours. With hard disk recorders, defragment your disk after each session. Get the dust out of those pots and faders too—a mixer cover is a great idea—and clean patch cords every four to six months.

Always remember that the arrangement is a critical element in the success of your production. A well-arranged piece of music can virtually mix itself if the tracks were recorded properly. On the other hand, a poorly arranged piece of music can be changed drastically by creative muting during the mixdown—that is, by "fixing it in the mix."

Don't worry about your gear's limitations. When it comes to recording, the most significant limitations are time, energy, and imagination. Where there's a vision and a will, there's nearly always a way.

Working alone in your studio provides great freedom, but the isolation can turn into a creative bottleneck. When you write, arrange, produce, and engineer everything yourself, you're not getting any new input. Working with other people, though it can be frustrating at times, often helps lift you out of creative ruts.

Recording Electronic Drums

Although electronic drums are instruments in their own right, for better or worse they are often compared to the real thing—and usually don't come out ahead. Let's look at some tips that will not only improve your recorded sound, but may improve the quality of the part as well.

Try triggering sounds and samples "live" rather than entering them in step time. If you're recording drum tracks in a sequencer, try to avoid using quantization. Play the part over and over, if you have to, until you get a feel that you like. Some drum modules let you use MIDI continuous controllers to make subtle changes in individual drum sounds. Used judiciously, controller modulation can definitely help. The track may not sound exactly like a real drummer, but it will sound less robotic.

Another approach is to buy a sample CD of drum loops, and create your part by looping the same rhythm over and over. Finding a loop that has the right tempo and the right style may not be easy, and once it's looped, it will sound almost as monotonous as a drum machine (however, you can always overdub some percussion manually—tambourine, maracas, shaker, etc.—to add interest). Some of these CDs provide samples of individual drums from the same kit that plays the loop. You can use these to fly in additional snare drum hits, tom fills, and so on that are different in every bar.

If you're not a drummer, you may be able to find one who is willing to play a part on MIDI pads. The natural accents, and the part itself, will probably be more realistic than anything you'd come up with using a keyboard as an input device. If you record the part into a sequencer, you can clean up occasional fluffed notes without disturbing the live feel, but be careful not to over-edit. Get the drummer to play the track several times if necessary, and then cut and paste the best bits into one finished track. Or, borrow a good snare drum and sample it yourself. Take five or six samples with as close to the same tone quality as you can manage, assign them to adjacent keys on the sampler, then alternate between the keys to add variety to the snare sound.

Limited Polyphony. Most drum machines have limited polyphony: if you try to play too many sounds at once, voices may be "stolen" (i.e., newer sounds will cut off the sustains of existing sounds).

One solution with multitracks is to sync the drums to a sync track, and record a few drums per track (mute the other drum sounds) on each pass. Or record some drums in your multitrack recorder, sync the drum machine to MIDI (or drive a drum tone module with a MIDI sequencer), and play the remaining sounds as "virtual" tracks during mixdown. But be careful: non-drummers frequently overplay parts, and have too many sounds going on at once. If you're running out of voices, that could be one of the warning signs of a bad arrangement.

One main voice-stealing problem is cymbals being cut off, as these have long sustains. If possible, overdub real cymbals and forget about using the electronic versions, which generally sound inferior anyway. In fact, you can often fool people into thinking you're using real drums just by overdubbing real cymbals.

Separate Outputs. In addition to stereo outputs, most drum modules have individual outputs, to which you can assign various drums. These outs are excellent for pulling the kick and snare out of the main mix and feeding them separately into the

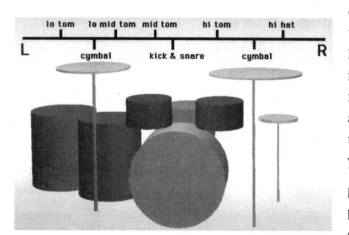

console. Typically, these two sounds are the foundation of a drum mix, and often benefit from custom EQ or other processing. It's also easier to change levels by moving a mixer fader than jumping into a drum module menu. Just remember to take any drums out of the main mix that you're feeding in separately.

❷ *Proper panning greatly influences the impact and realism of a drum kit during the mixing process.*

Panning. Decide if the listener is behind the drums or in the audience, as this influences where you position the hi-hats and toms in the stereo field. Some prefer the audience viewpoint, where toms are on the left and the hi-hat is on the right. Figure 2 shows a typical drum setup and where you might pan individual drums in the left/right stereo field; keep your hypothetical setup in mind as you set the panning.

Instead of panning drums full right and left, try closing them in a little bit toward the center. This leaves some room for percussion instruments, which tend to go more to the extremes. Arrange the panning so that drums don't fight with respect to frequency response, but complement each other. For example, with hi-hats on the right, place tambourines and shakers—which also have a lot of high frequencies—toward the left. Since toms are on the left, pan bongos, dumbeks, timbales, and other pitched drums to the right.

The Problem with Processing. Be careful when processing electronic drums. Most manufacturers already add EQ, treble enhancement, compression, etc., to make their drums as wonderful-sounding as possible. It's often better to seek out the dry drum sounds, and work with those so you can add your own processing.

A different reverb on each drum seems to diminish the power of the part. Instead, a combination of reverberation with short delays seems to work well.

Sequencing with Electronic Drums. If you want to avoid a mechanical feel, you need to think like a drummer. Here are some tips that should help.

Change Tempo. Real drummers speed up and slow down to add expressiveness to a song. These changes are often subtle, but have a huge effect on the emotional impact of the music. Your sequencer will have some option for varying the tempo—use it! Pull the tempo down one beat-per-minute just before going into the solo, push it one beat-per-minute higher during the solo, and so on.

Alternating Pitches. No two drum hits are exactly alike, but they are in the world of electronic drums. To get around that problem, try assigning the same drum sound to two different notes, and detune one very slightly compared to the other. Then use a "logical edit" function to move every other snare hit to the alternate drum sound. This is particularly helpful with drum rolls.

Track Shifting. Drummers often push or lag individual notes to create a particular "feel." For example, pushing the ride cymbal a bit ahead of the beat is a common jazz technique, whereas rock drummers often lag a little on the snare to create a "bigger" sound (subjectively, we naturally associate delay with distance). Sequencers generally include a track shift function that can move entire tracks forward or backward a certain number of MIDI clocks, which is ideal for this application. Others provide "groove templates," which can quantize drums to patterns other than traditional quantization templates. For example, these patterns might have been played by a real drummer or follow a particular type of rhythm (e.g., Salsa or Brazilian), thus imparting some of the feel that only a human drummer can truly provide.

Timing. Drum machines usually respond faster to MIDI data than do synths (particularly "workstation" types running in a polytimbral mode). To bring all sounds "into the pocket," consider shifting the drum track a few clocks later to compensate.

Velocity vs. Controller 7. To change drum levels when sequencing, you have two choices—alter velocity or (with some drum machines) feed MIDI Controller 7 messages into the drum module. Controller 7 usually sets the overall volume, whereas velocity affects individual drums. With multisampled drums, velocity changes may also affect the timbre. To change individual drum levels without altering timbre, do so at the machine itself, using the available drum level mix option.

Finally, remember to keep it simple. You'll seldom hear anyone say a drum part had too *few* notes!

Synchronization

Synchronization allows two devices to work in tandem by providing a precise master timing source, which these devices use as a reference. When two devices are synced, one is the *master* (the source of timing information, which it generates internally) and the other is the *slave,* which receives this timing data and reacts accordingly. Under these conditions the slave is in "external clock" or "external sync" mode, because it receives timing clock information from the master. In other situations, a dedicated master timing source may generate sync signals that all of the synchronized (slaved) devices receive.

For example, if you want to link two eight-track tape decks together for a total of 16 tracks, the tape transport mechanisms have to sync (lock) together. If they aren't synced, even if they start precisely at the same moment, sooner or later they will drift apart because of mechanical differences. This will create an echo effect of ever-increasing length as the devices slip further and further out of sync.

A common synchronization application for desktop musicians is syncing a sequencer to a digital multitrack tape recorder. Even if you're using a single computer to multi-task a software sequencer with a hard disk recording program, the two must still be synchronized. Computer-based devices usually have far more precise timing than motorized tape decks, but the problems due to any lack of synchronization only get more subtle; they don't go away.

Synchronization Applications. Following is a list of common applications for synchronization in the studio.

- Synchronize multitrack recorders together (whether analog or digital tape, hard disk, MIDI sequencing, or some combination) to create more tracks. Press the transport controls on one machine, and the others follow along. You can also synchronize a MIDI sequencer to tape or hard disk and record MIDI data in the sequencer, along with audio data in the hard disk or tape recorder.

- Synchronize an automated mixdown system and/or signal processing system to a multitrack recorder.

- Interdevice MIDI communication. For example, sync a drum machine via MIDI to a sequencer that is itself synchronized to tape or hard disk.

- Synchronize two or more devices to create a "multimedia" setup (e.g., sync an audio recorder to a video recorder to create a soundtrack).

- Digital audio synchronization. Digital audio signals need to be synchronized to a master clock to prevent certain types of "jitter" that can degrade sound quality.

Synchronization Signals. Different applications require different types of synchronization signals. The two main references for sync signals are *absolute* time (hours/minutes/seconds) and *relative* (musical) time (bars/beats). The two formats used for these are SMPTE and MIDI, respectively. First we'll cover the different kinds of MIDI sync signals, then SMPTE synchronization.

MIDI Sync

The MIDI Specification includes synchronization instructions in the form of digital timing messages, which are exchanged between pieces of equipment over the MIDI cable.

Start, Stop, and Continue. The most basic MIDI sync instructions are *start*, *stop*, and *continue* commands. These synchronization messages are examples of System Common data (i.e., all instruments on all channels, not just individual instruments on individual channels, receive this data).

- *Start* tells the slave when the master has started so that the two units can start together. Start always causes a device to start at the beginning.

- *Stop* tells the slave when the master has stopped so that the two units can stop together.

- *Continue* tells the slave to resume playing from where it was last stopped so that the two units can continue together from that point.

MIDI Clocks. Once you press Play, you now need signals that will keep the units in sync. The simplest option is the *MIDI clock* signal. A master device transmits clock bytes over a MIDI cable at a rate of 24 ppq (pulses per quarter note), so the master MIDI clock emits 24 MIDI clock messages every quarter note. When a slave receives one of these messages, it advances its internal clock by 1/24th of a quarter note.

In MIDI, clock messages have priority over all other messages to ensure accurate time keeping. Many devices, such as sequencers, further subdivide this clock rate internally so that events may be recorded with greater resolution (e.g., 1/480th of a quarter note).

MIDI clocks are tempo-dependent. If the master sequencer speeds up, it will transmit more MIDI clocks per second, and the slave will speed up too.

Song Position Pointer Messages. MIDI Song Position Pointer messages (SPP messages for short) identify precise locations in a song. They do this by keeping track of how many 16th notes have elapsed since the beginning of a tune (up to 16,384 total). An SPP message is usually issued just prior to a continue command, which provides autolocation. For example, suppose a drum machine is slaved to a sequencer. If you start the sequencer in the middle of a song (e.g., 724 16th notes into the song), as soon as you start the sequencer, the following events happen:

1. The sequencer issues an SPP message that tells the drum machine, "Hey, we're 724 16th notes into the song!"

2. The drum machine then autolocates to 724 16th notes from the beginning.

3. The sequencer pauses long enough for the drum machine to autolocate, then sends a continue message.

4. Both units play from that point on.

With SPP, you can record a sync track whose tempo varies; however, once the SPP data is recorded on tape (see FSK, below), the tempo is fixed and cannot be altered.

FSK. MIDI sync signals cannot be recorded directly as audio because they run at 31.25kHz—a higher frequency than multitracks designed for sound recording can handle. However, SPP data and clock pulses can be translated by hardware boxes into audio tones through a process called FSK (Frequency-Shift Keying). As you play back a sequence, the SPP data gets recorded on tape as an audio sync signal. On playback, these audio tones are decoded back into SPP messages; therefore the multitrack serves as a master MIDI clock that issues SPP messages.

Tempo Maps. What happens if you want the MIDI clock sync signal to change tempo at various points in the song? After all, a static tempo can be pretty uninteresting. The solution is a *tempo map*—a programmed series of tempo changes. Typically, you can change time signature on any measure, and tempo at any time.

Tempo maps require that the MIDI aspect of the song be structured in advance, so it's best to start composing with the MIDI sequencer. If the MIDI sequencer syncs to MTC (described next), it will follow the sync signal but add any tempo changes that you created. With SPP/FSK sync, the FSK tone recorded as audio reflects the tempo changes, and on playback, the sequencer follows these changes.

MIDI Time Code (MTC). This part of the MIDI specification allows SMPTE times (see "SMPTE Time Code," below) to be communicated directly over MIDI, thus allowing MIDI devices to respond to absolute times if necessary. An example will help get the point across of why this is desirable.

Suppose you're scoring a commercial, and also providing some sampled sound effects to be played back from a sampler. You sync your sequencer to the video, and start creating the background music. Then as you see where specific effects are to take place—a car crash, crowd applause, etc.—you play the corresponding keys on the sampler that trigger those sound effects, and record those keypresses into the sequencer. Perfect—on playback, the tune is in sync with the video, and all the effects are triggered in just the right places.

Then the producer decides to speed up the song tempo by five percent. So you speed up the sequence tempo, but now the rate at which the sound effects occur speeds up too, and they no longer match the film. The problem is simple: the sound effects relate to absolute time—in other words, the door should slam at perhaps 12 seconds and 11 frames into the film, regardless of what the music is doing. The music itself relates to relative (musical) time.

With a MIDI time code-equipped sampler, you could create an *event list* of SMPTE time code cues, then trigger the samples via MIDI time code at specific times. Problem solved—you can change the sequencer tempo, yet the MIDI time code values remain constant, since they relate directly back to SMPTE. Therefore, the samples will be triggered at the specified times regardless of what happens with the sequencer.

When a sequencer is synced to MTC, it references its own internal tempo—in the bars, beats, and measures of musical time—against the absolute time. For example, if the tempo is 120 beats per minute, the sequencer knows that the 60th beat has to line up with 30 seconds of elapsed time. A sequencer that responds to MTC can change tempo or time signature at any time; the sequencer simply re-calculates where notes should fall with respect to absolute time.

MIDI Synchronization in Action. Syncing MIDI gear to tape (or hard disk) is the equivalent of two multitrack recorders operating in tandem. You need not record drum machine or sequenced synthesizer parts, because the multitrack's master clock (or "sync track") can trigger the drum machine and/or sequencer in real time. Thus, the slave's audio outputs are virtually equivalent to the tape track outputs: they play along just as if they were recorded on a multitrack, and feed into your mixing console just like any other track. In fact, the outputs from the electronic instruments are often called "virtual tracks" because they are "virtually" the same as tracks recorded in any other type of multitrack recorder (see Figure 3).

Sequencers are also useful for controlling automated mixing and signal processing. The sequencer could be a "stand-alone" type, or part of an automated mixdown system that runs concurrently with other MIDI sequencers.

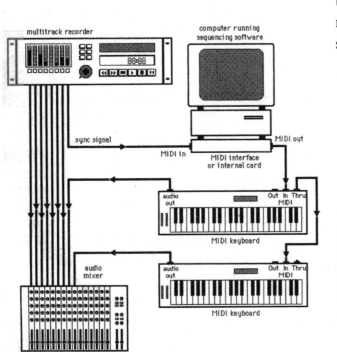

❸ Typical virtual tracks setup. The multitrack digital tape recorder provides a sync signal to which the sequencer synchronizes. The sequencer in turn drives two MIDI keyboards. As you start and stop the tape, the sync converter produces commands that start and stop the sequencer.

Not having to record sequenced synthesizer parts economizes on tape or hard disk tracks. Many project studios use a digital or analog multitrack to handle acoustic instruments, and virtual MIDI tracks to expand the total number of tracks.

Hard disk systems, because they are computer-based, can often add some useful twists. For example, a MIDI sequencer might be built into the program, so it's unnecessary to sync an external device. The hard disk system might also generate MTC or SMPTE time code (as discussed below) to provide synchronization signals.

SMPTE Time Code

The most common of the sync signals referenced to absolute time is called SMPTE, after the Society of Motion Picture and Television Engineers, which took a time logging protocol developed by NASA and adapted it to film and video applications. SMPTE divides time into hours, minutes, seconds, frames, subframes, and user bits. A number of frame rates are used to accommodate film and video, which run a different number of frames per second. When syncing to SMPTE time code, you'll have to choose matching frame rates—for example, 24, 25, or 30 fps (frames per second)— for the master and slave devices.

SMPTE time code is an audio signal, and can be easily recorded and played back. It can't be sent down a MIDI cable directly, as MIDI is a digital signal. However, MIDI Time Code (MTC), as mentioned above, allows SMPTE data to be encoded in MIDI form and transmitted over a MIDI cable. Note that SMPTE and MTC don't provide start or stop commands, nor do they change tempo. They provide an absolute timing reference in minutes and seconds, rather than a music-related reference in bars and beats. If a sequencer is synced to MTC and you change its internal tempo *after* recording some tape tracks, it will still drift out of sync with those tracks, even if it's receiving MTC correctly from a SMPTE track that has been recorded ("striped") onto the tape. This is because the tempo is independent of the absolute time expressed by MTC or SMPTE.

SMPTE time code is a reliable, and increasingly universal, method of synchronization. With tape, if some slight damage occurs to the tape so that SMPTE data is lost, most SMPTE boxes will "guess" where the markers would have been until the real markers appear again. In contrast to Song Pointer-based tape sync, SMPTE is a standard. SPP boxes have their own proprietary way of translating SPP messages into audio tones, so you usually have to use the same device for recording and playback.

How SMPTE Works. A *SMPTE time code generator* generates "timing markers" that serve as a super-accurate index counter. Hard disk recording systems often generate SMPTE timing messages without requiring external hardware or a dedicated sync track. Sync control units for digital tape machines, such as the Alesis BRC, can also provide SMPTE messages.

A frame's duration varies for different applications. For film work, the standard rate is 24 frames per second (i.e., each frame equals 1/24th of a second). For black and white video, the rate is 30 frames per second in the U.S. and 25 frames per second in Europe. For U.S.A.-standard NTSC color, the rate is 29.97 frames per second.

A *SMPTE time code reader* reads the SMPTE markers, and sends this timing data to other devices.

SMPTE-to-MIDI Conversion. A SMPTE-to-MIDI converter can convert absolute time (as represented by SMPTE) into "musical time" by reading SMPTE, then translating SMPTE times into MIDI SPP data. Athough common at one point, these boxes are far less used today because most MIDI seqencers can follow MTC or SMPTE directly.

Advanced SMPTE-to-MIDI boxes accommodate tempo changes by letting you define a tempo map, where every time you enter or tap a beat, it is related to SMPTE time code (e.g., at 0 hours, 2 minutes, and 12 frames, increase the tempo by 2 beats per minute). On playback, the SMPTE time code provides a consistent reference point for the beat map. Many jingle and film scorers use the beat map feature to "fudge" the tempo a bit in spots so that particular visual cues, like cereal pouring out of a box, Godzilla eating Manhattan, etc., land exactly on the beat. In other words, if the cue lags a bit behind the beat, the tempo will subtly speed up for a couple of measures prior to the cue.

Many computer interfaces have SMPTE-to-MIDI conversion built in, which simplifies matters. Feed the SMPTE sync track audio signal into the interface's SMPTE audio input, and SPP messages magically flow out the interface's MIDI out jack. In today's studio, SMPTE generally provides the system master clock, while MIDI acts as an intelligent interface between various pieces of equipment.

Word Clock

Neither MIDI nor SMPTE provides enough accuracy for synchronization in digital audio recording. High-end digital audio devices are locked together using a signal called *word clock*. This clock signal is accurate to a single sample word—usually 44.1 or 48kHz. Specialized hardware is required for syncing a digital audio recorder's word clock to SMPTE or some other timing reference. For more on word clock, see "Digital to Digital Transfers," below.

Quantization

Quantization, which shifts the start times of MIDI notes to a specific rhythmic "grid" (eighth notes, 16th notes, etc), got a bad rap back in the days of 24 ppq sequencers. If you didn't quantize, the clock resolution was so coarse that what came out of the computer was a little jerky compared to what you had played, so you quantized everything, and then the music marched along in lockstep. Lockstep is a wonderful thing for certain styles—but what if you want your tune to have that supple human feel?

One school of thought is, don't quantize anything. Play the part ten times if you have to, until you get it right. Again, for certain styles—classical orchestra simulations, say—that's unquestionably the right approach. But for pop music, consider using a mixed strategy.

Start by quantizing the kick and snare so that they're nailed to the beat. That provides a fixed reference point for the listener, and for the other tracks to be recorded. What happens next depends on the music. Consider using percentage quantization to move the bass line 50% of the way to the quantization grid, but it might be better to quantize the bass all the way when it's playing a purely supportive role, then use a 50% setting, or no quantization at all, when the bass steps forward to play a more melodic line. Ditto for chord comps and hi-hat patterns.

Stay away from the "human feel" utility, by the way. This generally randomizes the note start times—but humans don't play randomly. Inspect your own live-recorded tracks (or those of a world-class player, the next time one drops by) and you'll find expressive *non*-random fluctuations. A real player will get excited, for instance, and jump on a fill a little ahead of the beat. Some players consistently play ahead of or behind the beat, or play certain licks such as grace-notes with a specific type of rhythmic displacement.

For the same reason, it doesn't make sense to quantize note durations. If you move the start time of the note by quantizing it, you may want to use the pencil tool to shorten or lengthen it slightly so that it ends exactly where it did before. Phrasing has a lot to do with the ends of the notes.

Hand Work. Another approach, which takes more effort but gives very musical results, is to leave a track unquantized and correct individual notes by hand. Open up the "piano-roll" notation window (see Figure 4), zoom in far enough to see clearly which notes are before or after the beat, and start playback. If either your ears or your eyes tell you that a note is too far away from the beat, stop playback, grab the note with the mouse, then rewind a couple of bars and start playback again.

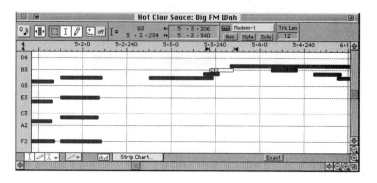

❹ In the piano-roll edit window in Opcode Studio Vision, the note that starts just before bar 5, beat 3, clock 240 has been selected and is being dragged to the right to bring it a bit closer to the beat. The mouse cursor in Studio Vision turns into the symbols <+> when this edit is being performed.

When using this technique on fast triplets and such, it's often helpful to shorten some notes and lengthen others slightly while you're at it. This will give a smoother legato, or the right sort of detached articulation, depending on what you want. If you're going to go to all this trouble, you may as well leave the velocity window open and fix any notes that are too loud or soft.

The big fill leading into a dramatic downbeat is often the place for a little extra hand work, even on a previously quantized part. Pushing the kick or tom note back so that it's a little late will add drama—and you might want to shift the crash cymbal so that it's a little early, even if you played it smack on the bar line.

Swing and Percentage Quantization. Two of the more common forms of quantization are *swing* and *percentage* quantization. When swing is applied, notes that fall in the second half of each beat are delayed by the swing amount, producing the dotted or triplet rhythm characteristic of jazz. In percentage quantization, notes are moved only part way to the chosen rhythmic grid: How far they are moved is determined by the quantization percentage or *strength* setting. For example, if a note is originally recorded 10 clock ticks before the beat, and the quantization strength is 50%, after quantization it will play back five ticks before the beat (see Figure 5).

You can also quantize one track to match another. The reason for using this feature would be if you've crafted a rhythm—on a hi-hat, let's say—in which certain beats are pushed or laid back more than others. Even then, it's probably best to let that part occupy its own place in the mix by *not* cluttering up its attacks with exactly matching hits on other drums.

5 Vision's quantization option in action. Measure 6 shows notes as originally played. Measure 7 shows the same notes quantized with 50% strength to the nearest quarter-note. The notes are closer to the beat, but not quite there yet. The notes in measure 8 have been quantized according to the instructions in the Quantize dialog box on the right (100% strength, quarter notes).

Looping—Steinberg's ReCycle

Steinberg's ReCycle is a cross-platform (Mac or Windows) program for musicians who use sampled rhythm loops of any kind—drum loops, percussion accents, bass lines, etc. It doesn't actually do anything you couldn't do manually in a sample editor (or, for the really bold, from the front panel of your sampler), but it does it a whole lot faster.

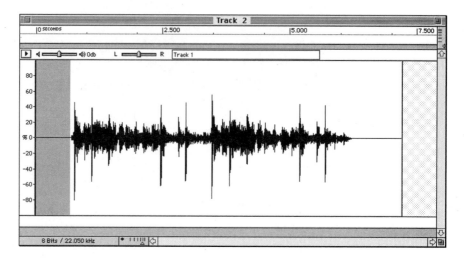

G In this two-measure rhythm pattern, the unnecessary dead space before the downbeat has been selected for trimming. When selecting, be careful to stay clear of the downbeat's attack—if you trim it off, the sound may click or lose punch.

ReCycle sets extremely accurate loop start and end points, and it makes it easy to integrate your sample loop into a MIDI sequence (see Figure 6). It also allows changing a loop's tempo without changing pitch, yet doesn't use time compression/expansion routines, which often produce undesirable digital artifacts. The program is easy to learn, so you can get results almost instantly.

Here's how simple the process is: Import any monophonic, 16-bit, 44.1kHz sample file and then click a button to play the file back. As it plays, move an onscreen "sensitivity" fader—ReCycle then uses an intelligent algorithm that searches for transients in order to locate each of the individual sonic components within the file. If the sample file is a drum track, each of these components (called "slices") will be an individual drum hit; if it is a bass track, each component will be an individual note. To preview the sound contained within each slice, simply click on it (the Mac version of ReCycle supports all Sound Manager 3.1-compatible hardware, such as the Digidesign cards, so you can preview in full 16-bit mode). Left-right markers appear at the file's start and end, and can be snapped to any slice start point, making it incredibly easy to set extremely accurate loop start and end points.

If ReCycle finds some false slice points (for example, in a drum flam, it may find both hits), individual slice points can be hidden. Conversely, if it misses some slice points (as it may do at lower sensitivities), you can insert them manually. Zoom in and out tools make the process easy, though it could be made even easier if some kind of audio scrubbing tool were available as well.

Next, tell ReCycle how many bars and beats the loop contains—it automatically calculates the loop's tempo. At this point, you can either download the individual slices between the left and right markers to your sampler (assigned one to a key; currently supported samplers include Digidesign's SampleCell and various Akai models, Kurzweil K2000 and K2500, E-mu, Ensoniq, and Roland samplers) or you can export them in a number of standard file formats (SDI, SDII, or AIFF in the Mac version; .WAV or AIFF in the Windows version). These samples can optionally be gain normalized to insure that each sample uses the maximum available headroom, but this also alters the dynamic balance between samples. At the same time, you can instruct ReCycle to generate a MIDI file that will play back the sample slices, one after the other, at the original tempo.

ReCycle also lets you specify a new tempo for loop playback, and the MIDI file the program generated will follow that new tempo. The magic of all this is that, by breaking a loop up into its individual components and triggering each component from a discrete MIDI note, you can alter the loop's tempo without changing its pitch.

One potential problem is that if you slow down the loop's tempo, there will be silences after each slice plays back. These silences may be apparent, especially if there is ambient sound in the sample. ReCycle cleverly addresses this problem through the use of a "stretch" function that adds short backward-forward loops to the end of each slice, thus extending each sample by a certain user-defined percentage. Since the end of a slice most likely contains only ambient sound (or, at worst, the tail-off of the sound), this works remarkably well.

It's not even so much what ReCycle does, but what it allows you to do after the fact that makes the program so impressive—the fun really starts after you've created and downloaded (or exported) all the individual slices and the MIDI file. For example, if your sampler provides multiple hardware outputs, you can route each component to a different output for additional mixing control, even applying different effects processing to each sample. Get bizarre by reversing playback of individual components (nothing like adding a few backwards snare or cymbal hits to an otherwise dull drum loop!) or experiment by substituting different sounds—simply assign a different sample to individual key numbers. In your MIDI sequencer, you can, of course, change tempo to your heart's content, and, if you're using Steinberg's Cubase Audio (or another sequencer that supports "groove" templates), you can even quantize MIDI tracks to the feel of the sampled loop.

ReCycle ships with an audio CD of demo loops as well as a concise owner's manual that contains a brief tutorial—the Macintosh version also includes drivers and extensions that support any Digidesign sound card. Stereo files are not yet supported.

An oddity worth mentioning is that unlike virtually every other piece of software in existence, ReCycle doesn't allow you to save your work, the thinking presumably being that once you've downloaded or exported your data, you're done. But I could imagine scenarios in which the user might want to go back and tweak some slice points or set different loop start/end points, so I do think this is an omission of note, though not necessarily a serious problem.

All in all, ReCycle is a program that successfully combines stunning simplicity with incredible utility. If you work with sampled rhythm loops, do yourself a favor and check it out—it will not only save you valuable time but can also lead you to new creative outposts.

Compression

Compressors are some of the most used, and most misunderstood, signal processors. While people use them to make a recording "punchier," they often end up dulling the sound instead because the controls aren't set optimally.

What It Does. Compression was originally invented to shoehorn the dynamics of live music (which can exceed 100dB) into the restricted dynamic range of radio and TV broadcasts (around 40–50dB), vinyl (50–60dB), and tape (40–105dB, depending on type, speed, and noise reduction used). As shown in Figure 7, this process lowers only the peaks of signals while leaving lower levels unchanged. A second process—which could be a gain-normalization utility with software compressors, or an output volume knob on a hardware unit—then boosts the overall level to bring the signal peaks back up to maximum. (Bringing up the level also brings up any noise as well, but you can't have everything.)

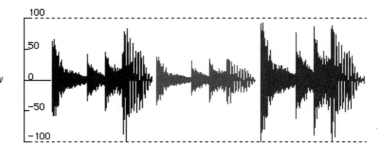

7 The first section shows the original audio. The middle section shows the same audio after compression. The third section shows the same audio after compression and turning up the output control. Note how softer parts of the first section have much higher levels in the third section, yet the peak values are the same.

Even though media such as the CD have a decent dynamic range, people are accustomed to compressed sound. Compression is also useful to help soft signals overcome the ambient noise in typical home listening environments.

Compression can create greater apparent loudness. (Commercials on TV sound so much louder than the programs because they are compressed without mercy.) And of course, compression can smooth out a sound's dynamics—from increasing piano sustain to compensating for a singer's poor mic technique.

How It Works. Compression is often misapplied because of the way we hear. Our ear/brain combination can differentiate between very fine pitch changes, but not amplitude. So, there is a tendency to overcompress until you can "hear the effect," giving an unnatural sound. Until you've trained your ears to recognize subtle amounts of compression, keep an eye on the meters. You may be surprised to find that even with 6dB of compression, you don't hear much apparent difference—but bypass the unit, and the difference will be obvious.

Compressors, whether software- or hardware-based, have these general controls:

Threshold sets the level at which compression begins. Above this level, the output increases at a lesser rate than the corresponding input change. Bottom line: With lower thresholds, more of the signal gets compressed.

Ratio defines how much, when the input signal rises above the threshold, the output signal level changes for a given change in the input level. For example, with 2:1 compression, a 2dB increase at the input (assuming that the input is higher than the threshold) yields a 1dB increase at the output. With 4:1 compression, a 16dB increase at the input gives a 4dB increase at the output. With "infinite" compression, the output remains constant no matter how much you pump up the input. Bottom line: Higher ratios increase the effect of the compression. Figure 8 shows how input, output, ratio, and threshold relate.

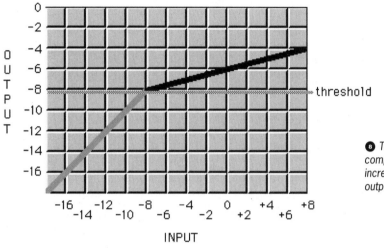

threshold

⑧ *The threshold is set at -8dB, and the compression ratio is 4:1. If the input increases by 8dB (e.g., from -8 to 0), the output only increases by 2dB (from -8 to -6).*

Attack determines how long it takes for the compression to start once it senses an input level change. Bottom line: Longer attack times let more of a signal's natural dynamics through, but those signals are not being compressed. In the days of analog recording, the tape would absorb any overload caused by sudden transients. With digital recording, those transients clip as soon as they exceed 0 VU. With short attack transients, this may not produce any significant audible degradation (that sound you hear is thousands of mastering engineers recoiling in horror). If there is distortion, lower the overall level with the...

Output control. Since we're squashing peaks, we're actually reducing the overall peak level. Increasing the output compensates for the volume drop. Turn this control up until the compressed signal's peak levels match the peak levels of the bypassed signal.

Decay sets the time required for the compressor to give up its hold on the signal once the input falls below the threshold. Short settings are great for special effects, like those psychedelic '60s drum sounds where hitting the cymbal would create a giant sucking sound on the whole kit. Longer settings work well with program material, since the level changes are more gradual and produce a less noticeable effect.

The *hard knee/soft knee* option controls how rapidly the compression kicks in. With soft knee, when the input exceeds the threshold, the compression ratio is less at first, then increases up to the specified ratio as the input increases. With hard knee (as illustrated in Figure 8), as soon as the input signal crosses the threshold, it's subject to the full amount of compression. Bottom line: Use hard when you want to clamp levels down tight (for instance, to prevent clipping in a power amp), and soft when you want a gentler compression effect.

Side chain jacks let you insert filters in the compressor's feedback loop to restrict compression to a specific frequency range. For example, inserting a high-pass filter compresses only the high frequencies—perfect for de-essing vocals.

The *link* switch in stereo compressors switches the mode of operation from dual mono to stereo. Linking the two channels together allows changes in one channel to affect the other channel, which is necessary to preserve the stereo image.

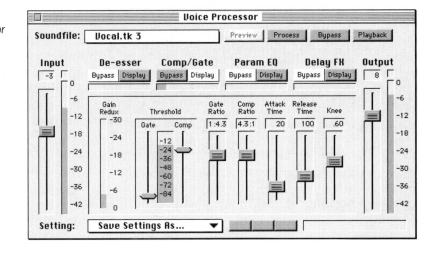

❾ *The compressor module from the Jupiter Voice Processor (Macintosh).*

Figure 9 shows the compression module from the Jupiter Voice Processor, a software plug-in for Sound Designer II running on the Mac. This is a typical setting used for vocals. There was close to −6dB of noise reduction at the moment this screen shot was taken, as shown by the Gain Redux meter. There are the expected threshold, ratio, attack, release, knee amount, and output controls; this module also includes a noise gate section with its own threshold and ratio controls.

Compressor Types: Thumbnail Descriptions

Compressors now come in both hardware varieties (usually a rackmount design) and as software "plug-ins" for existing digital audio-based programs. Following is a description of various compressor types.

- *Old faithful.* Whether rackmount or software-based, typical features include two channels with gain reduction amount meters that show how much your signal is being compressed. Hot tip: When compressing a stereo mix, blend in some dry, non-compressed signals for a livelier, more dynamic sound.

- *Multiband compressors.* These divide the audio spectrum into multiple bands, with each one compressed individually. This allows for a less "effected" sound (for example, low frequencies don't end up compressing high frequencies), and some models let you compress only the frequency ranges that need to be compressed.

- *Octal compressors.* These house eight compressors in a single rack space, which helps reduce the possibility of overload for signals going into your recorder, as well as brings back some of the tape compression effects associated with analog tape (minus the distortion). These units are overkill if you're overdubbing tracks one or two at a time (just get a good stereo compressor), but for live recording, they can literally save a session.

- *Vintage and specialty compressors.* Some swear that only the compressor in an SSL console will do the job. Others find the ultimate squeeze to be a big bucks tube compressor, like the Demeter or Groove Tubes models. And some guitarists can't live without their Dan Armstrong Orange Squeezer, held by many to be the finest budget sustain box ever made. Fact is, all compressors have a distinctive sound, and what might work for one sound source might not work for another.

Whatever kind of audio work you do, though, there's a compressor somewhere in your future. Just don't overcompress—in fact, avoid using compression as a cop-out for bad mic technique or dead strings on a guitar. It is an effect that needs to be used subtly to do its best.

Backing up Your Data

Your floppy disk containing those killer sequences, the DAT master tape for your next CD, that S-VHS or Hi-8 tape that's backing up your precious multitracks where Trent Reznor came in and played a few guest licks—they're all going to be history, sooner or later.

The question is: How soon is soon? The answer: Sooner than you think, which is why backing up is crucial. This is particularly important with digital tape; unless your machine has an error indicator that chronicles the tape's gradual breakdown, you'll have no clue that it's about to become unuseable. What's worse, unlike analog tape (where minor dropouts can often be tolerated), a minor dropout with digital can produce a rude, ripping sound.

Which backup method is the most reliable? Before getting to that, let's solve some problems before they start.

An Ounce of Prevention. Several steps can help increase the shelf life of digital tape, whether it's DAT or multitrack.

- Clean and maintain your machine. Use a dry head cleaner (e.g., Maxell's version for DAT) for periodic cleaning; every 20–30 hours seems about right. This should take at least the major pieces of gunk out of the tape path.

- Under normal use, have your DAT or multitrack professionally maintained once a year (for heavy use, every six months). Slight misalignments of the head blocks and other components are a major reason why tapes get "eaten."

- Use high-quality tape. Not all tape batches are created equal. With DAT, don't trust your precious music to recycled computer tapes. The latter are often thinner than audio tapes, which makes for a bad match with the transport. All things being equal, thicker tapes are more reliable than thinner tapes.

- Pre-condition your tapes. Fast-forward and rewind several times before use (for ADAT, get a separate VHS tape rewinder so that any gunk that gets shed goes into the rewinder, not your tape machine). This unpacks the tape properly.

- Eject the tape during a silent section, as the greatest amount of tape handling inside the machine happens during insertion and ejection.

- Avoid cue and review modes when rewinding, as the tape is under more tension with respect to the head. Use the normal rewind and fast-forward functions.

- Store your tapes (and floppy disks) properly. Avoid excessive humidity, heat, and cold. Safe deposit boxes and climate-controlled office buildings are good options. Also, it's generally good practice to play a tape through to the end before storing (a holdover from analog days, but it applies to digital tape as well). This precaution makes sense because the tape will be packed more evenly on the hub after normal-speed play than after a fast-forward or rewind, which could produce areas where the tape is packed more tightly than normal.

One more point: Any magnetic media can be erased if stored near strong magnetic fields. Never put tapes or floppies on top of speakers, near transformers, etc. When shipping a tape, pack it in the middle of a relatively large box just in case it gets placed somewhere inappropriate during transport.

Backup Options. So what's the longest-lived form of backup? Sadly, if there's a fire or other physical damage, nothing works. This is why remote backups, such as a safe deposit box, are always prudent. Here are the options:

DAT was never intended as a robust, professional medium; the tape is thin and the tracks themselves are $1/10$ the thickness of a human hair (are you nervous yet?). You can count on a shelf life of at least five years if the tape was good stock to begin with, and if you've been careful about handling and tape storage. Otherwise, you could start getting errors much sooner than that.

S-VHS tapes, as used in ADAT, tend to be somewhat less delicate than DAT. Still, there remains the issue of magnetic and plastic decay. Look at videotapes from 10 years ago: Those occasional streaks and "snow" might be a minor annoyance when watching *The Empire Strikes Back*, but if that was digital audio and the machine's error correction couldn't compensate, the cut would be toast.

Removable hard drives (SyQuest, Iomega, etc.), although theoretically more reliable than DAT, are still magnetic media. I've heard plenty of horror stories about cartridge reliability, but to be fair, these often come from people who don't keep the cartridges in their sealed cases, and throw them in the back seat of a car ("removable" doesn't mean "transport with impunity"). Cartridges are also fairly pricey, but there's another issue: If you pull a disk out of the archives in 10 years, will there be a working mechanism on which you can play it back? The multiplicity of removable cartridge formats is in itself a problem.

Magneto-optical cartridges boast very high reliability, immunity to magnetic fields, and large amounts of storage (1GB drives are commonly available). MO cartridges are the preferred backup medium if you need near-hard disk speed, transportability, and robustness.

A *CD-ROM recorder* is an excellent option. These models need to be used with computers; whatever you back up has to be stored as a digital audio file, then transferred (usually through SCSI) to the drive. Budget CD-Rs that simply record audio are another option. However, a CD-ROM writer lets you back up computer data, as well as make one-off CDs.

Initially, when a CD recorder wrote data to disk, that section of the disc could not be reused. Later models can use eraseable media, which although more costly, are more environmentally "friendly" as they can be reused. There is one caution: most CD recorders are optimized to work with a particular type of recordable CD formulation. Find out which type your manufacturer recommends, and stick with that type.

Backup Your Backups. Even CDs aren't perfect; no one really knows how long recordable CDs will last. Although some studies have said 70 years, exposing CDs to light or heat can shorten that dramatically, and eraseable CDs are estimated to last no more than 25 years. Since no backup system is perfect, it's a good idea to do the following:

- Save to multiple media. For example, back up DATs to two tracks of an ADAT tape, and if you have a CD recorder, back up to that too. It's a good idea to back up really important data to multiple tapes of the same type. That way, if there's a problem with a tape, you can transfer the existing data into a hard disk recording system, and use an alternate tape to fill in the missing section(s).

- Re-backup periodically. As soon as you buy a tape, date it. After three to five years (depending on your level of paranoia), back up to a fresh tape. It's digital, after all, so you won't get any deterioration when you make copies (unless the number of dropouts is so severe that the error correction has to "fill in the blanks"). Even if you don't have digital ins and outs, at least back up through your DAT's analog I/O. There will be a slight sonic deterioration, but that's better than losing a piece of music altogether.

If you have only one DAT deck, you can re-backup by borrowing a DAT from a friend, or by digitally transferring over to a computer running a digital recording program, then sending the file back to a fresh tape. Use a consistent sample rate when backing up.

The bottom line is that an optical medium beats tape, but if tape is all you can afford, make multiple backups—and whichever medium you use, keep your fingers crossed.

Digital-to-Digital Transfers

One of the great aspects of the digital revolution is that we can now "clone" digital audio, transferring it from one medium to another without the deterioration associated with analog copying. You can blast signals onto digital tape, send them over to a hard disk system for editing, then transfer the best bits over to a sampler for triggered playback from a keyboard—all without degrading the sound. Try doing that with analog gear!

Although digital transfers are generally straightforward, there are still a few complications along the way—so here are some hints, tips, and cautions.

Transfer Protocols. Currently there are several common hardware/software digital transfer protocols that you may encounter:

- AES/EBU. This protocol—the acronym stands for "Audio Engineering Society/European Broadcast Union," the two groups that collaborated on defining it—is a two-channel digital audio signal designed for professional applications. Signals are carried over wire cables that terminate in XLR (3-pin) connectors, or via fiber-optic cables. AES/EBU inputs ignore SCMS copy protection data. (SCMS, the Serial Copy Management System, is a misguided attempt to prevent copyright infringement. It prevents making a digital copy of a digital copy.)

- S/PDIF. This consumer-oriented version of the AES/EBU protocol was first adopted by Sony and Philips, hence the acronym, which stands for "Sony/Philips Digital Interface." S/PDIF (pronounced "spih-diff") uses RCA phono jacks or optical (TOSLINK) connectors; SCMS encoding is generally recognized, although some DAT decks and other devices with S/PDIF connectors allow you to defeat SCMS.

- MADI. Found in a few pro-level multichannel digital decks, a MADI fiber-optic cable can transfer up to 56 tracks of digital audio. The acronym stands for "Multichannel Audio Digital Interface."

- SDS. The world's slowest way to get digital audio from one place to another, SDS (the MIDI Sample Dump Standard) passes digital audio down a MIDI cable. It is used only with samplers.

- SMDI. This successor to SDS sends digital audio over the SCSI bus and associated control data over a MIDI cable, achieving a 50x increase in speed compared to SDS. The acronym stands for "SCSI Musical Data Interchange." Neither SDS nor SMDI is used much with standard digital recording.

- Proprietary. Not all recording devices follow a particular standard (as the old saying goes, "People must like standards…there are so many of them"). Instead, many of them implement their own proprietary transfer method. For example, the Alesis ADAT can transfer eight channels of digital audio over a fiber-optic cable; this has become such a common standard that other companies make ADAT-compatible gear (including mixers, synthesizers, and hard disk recorders) that can communicate using the same protocol. In fact, the ADAT Optical Interface has become a de facto multichannel digital audio interface.

Digital hookups are not that different from analog ones—outputs go to inputs—with one exception: These signals run at a much higher frequency than standard audio. Fiber-optic cables can handle this, but conventional wire cables should be low capacitance. In many cases, standard audio cables are not satisfactory because they distort the waveform, which may cause jitter (see below).

The Last Word (Clock). The numbers that make up digital audio have to arrive at the right place at the right time, which is a job for synchronization. "Word clock" refers to the timing (sampling) rate, typically 44.1 or 48kHz. A clock running at this rate provides the common timing reference between two interconnected digital audio devices.

For example, assume you want to edit two tracks of DAT audio on a hard disk editing system such as Sound Forge. When bouncing signals from DAT to Sound Forge, the DAT generates the word clock, so Sound Forge needs to listen to this external clock source. In fact, it derives its word clock by observing the actual timing with which sample words arrive at the AES/EBU or S/PDIF input.

When bouncing from Sound Forge back to DAT, Sound Forge listens to its internal clock. The DAT, being in record mode, recognizes the clock coming in at its digital input and syncs to that. (Had Sound Forge been set to external clock when transmitting, the transfer would fail, because both units would be looking for clock signals to sync to, and not finding them.)

Note that feedback loops are possible with digital audio. If you encounter feedback, simply disconnect the cable that's not being used. For instance, if you're transferring from the DAT's digital out to the computer's digital in, disconnect the cable that goes from the computer's digital out to the DAT's digital in.

Level Setting. Digital-to-digital transfers require no level-setting. Sixteen bits of "full code" (maximum level) is 16 bits of maximum level, period; the same goes for 20- and 24-bit data. Don't worry too much if your DAT flashes the "over" indicator when it receives audio that is obviously not clipped. This simply means that the indicator triggers just *before* the signal goes into clipping.

The Clone Controversy. Some "golden ear" types insist that different digital clones can sound different. But digital just doesn't work that way; a stream of digits is a stream of digits.

If a few digits are lost somewhere along the line, error correction methods will kick in to replace them. But only in extreme cases (such as large tape dropouts) will this become audible. Some digital data formats, such as CDs, include redundant data or allow missing data to be re-read from the source medium, so that an error can be corrected with 100% mathematical exactness. As long as the transfer process is free of errors, you can clone digital data all day: Compare the bitstreams of the original and the copy, and they'll be identical.

Digital Processing During Transfers. There are times, such as when a mastering house loads up your DAT master prior to pressing your new CD, that digital EQ or other processing may be patched in during the transfer. In this case, we're no longer dealing with a clone.

Many processors claim "24-bit internal processing" even though the A/D and D/A converters at the analog input and output are 16-bit. The reason for the extra bits is because gain changes within the unit (filtering, compression, etc.) cause numbers to be multiplied and divided, which can create values that require a longer word length than the usual 16 bits. (To take a simple example, if you multiply 3 x 6 and then divide by 2, you only need one digit for the inputs [3 and 6] and for the result [9], but you'll get an incorrect answer unless you can internally store the result of the multiplication [18] as a two-digit number.) Twenty-four-bit devices can handle the extra math overhead without too much rounding off, but the more signal processing you add and the more that signals are truncated back to 16 bits for transfer to another device, the more errors accumulate. These affect mostly the very low end of the dynamic range, but anyone who has cringed at the sound of a non-dithered reverb tail breaking up into crunching during the final stages of decay knows that low-level artifacts can be a problem.

Sample Rate Conversion. Sample rate conversions face several technical problems, so it's best to choose a consistent sample rate throughout the recording chain. Fans of the "ADAT classic," take note: Although you can set the sample rate to 44.1kHz, this works by slowing down the overall system clock. If you're synching a sequencer with the BRC, the tempo will come out slower by a factor of 44.1/48. Either increase the sequencer tempo by 48/44.1 (1.08843), or just keep your system at 48kHz. Then, to achieve a final sample rate of 44.1kHz on the DAT master, mix down the ADAT signal through an analog mixer and set the DAT to record at 44.1kHz. Think of it as "analog sample rate conversion."

Mixing

5

Basics of Audio Mixing

You know that part in every good suspense thriller where the persevering hero unlocks the mystery, and all the characters, clues, twists, and turns fall perfectly into place? For the recording musician, mixdown—the last stage of the recording process—serves the same purpose. In a good mixdown, the jumble of musical parts comes together; sonic textures are sculpted, all the elements are blended, and what's left is a unique musical tale, played out on a unique soundstage.

Fortunately, giving your mixdowns a full, balanced, and professional-quality soundstage needn't be a big mystery.

What's a Mixdown? The song is written, the performances are captured on disk, sequencer, or tape, and the work is ready to come together. Mixdown is the process of bringing together separate tracks into a unified whole. The tracks can be audio tracks (from hard disk or tape), MIDI tracks (from a soundcard, tone generator, or digital keyboard), or a combination of both. Usually, at least three or more such *source* tracks are mixed down to the left and right audio tracks of a stereo *mastering deck,* such as a DAT recorder or recordable CD (see Figure 1).

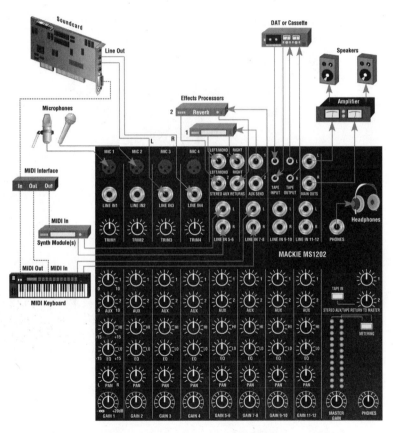

❶ The mixer is the studio's "traffic director": it routes signals from various inputs to various outputs.

Before mixdown, your tracks are like pieces of a jigsaw puzzle spread out on the table: You can examine them one by one, and you even throw a few of them together loosely to get an idea of the big picture. But it's only during mixdown that you decide which piece goes where, so that your final mix—your master tape—has all the pieces locked in place.

The usual purpose of a mixdown is to create a "portable" master recording, which, for instance, can be played in a car or sent to a friend, record label, or professional duplication house. But a stereo master tape isn't the only fruit a mixdown can bear. Say you're creating a soundtrack for a QuickTime movie or Macromedia Director presentation; in this case, you may be mixing directly to your computer's hard disk, and you may prefer to do so in mono—especially if you consider that CD-quality audio uses approximately 5MB per minute of hard disk space in mono, as opposed to 10MB per minute in stereo.

What Happens During Mixdown? Mixdown can be the most creative—and decision-intensive—step of the recording process. It's where you determine the sonic qualities of each track, such as:

- Relative level, from loud to quiet, or sometimes silent (muted)

- Overall tone, using tools such as equalizers (EQs) and synthesizer filters or brightness controls

- Location within the stereo panorama, from left to right

- Location in terms of "depth" and "room size" (from near to far), using reverb, delay, and other effects processing.

In essence, a mixdown is your opportunity to create a virtual soundstage. When you adjust levels, you're the conductor. As you adjust tone, you get to be an instrument designer. When you localize tracks in terms of depth and room size, you become an acoustical architect. And when you position the instruments from left to right, you place the musicians on your virtual soundstage.

Mixdown is also the last opportunity to correct problems that cropped up in the original recording, hence the phrase, "We'll fix it in the mix." For example, it's common practice in professional recording studios to create composite vocal tracks by stringing together the best phrases—even *syllables*—from multiple vocal performances. Whether this sacrifices the flow of the performance in a quest for perfection is something only you can decide.

Incidentally, although mixing in a professional studio can involve scads of gear, with 48 or more audio tracks routed through gigantic mixing consoles (and towering racks of outboard signal processors) on their way to the mixdown deck, the principles of mixdown remain the same whether you're using a half-million-dollar console or simply a MIDI sequencer. Using just a computer and sequencing program, a synth (or a high-quality synth-equipped soundcard), and a DAT deck, it's possible to craft surprisingly good-sounding mixes. The keys are your imagination, listening talents, and experience.

By the way, when we refer to mixdown as "decision-intensive," don't let that sound intimidating or even final. Most hits you hear on the radio were mixed after days, if not weeks, of experimentation. If you don't like the results, you can always go back and mix it again. If anything, mixdown can be the most fun part of recording. After all, the musical performances are already captured, so the pressure to "perform" is off; a wealth of cool signal processing tricks is at your disposal, and what was previously rough and unappealing can be polished to a fine sheen (assuming, musically speaking, that's what you're after).

Analyze Your CDs. Before you touch a fader or grab a mouse, listen closely to some favorite CDs, both on speakers and headphones. As you mix your own music, switch back and forth between it and CDs of a similar musical style. How loud are the drums in relation to each other and to the rest of the instruments? How about the vocal or solo instruments? Strings? Brass? Sound effects? Note the balance of frequencies—a common mixing mistake is cranking up the high and/or low end, but too little of either is just as bad as too much, leading to dull or wimpy mixes. Close your eyes and try to pick out each instrument. Where is it in the room? Does it echo or reverberate? Does it seem to spread out beyond the speakers? Do instruments in the rhythm section (particularly bass and drums, but also piano and rhythm guitar) appear to be in the same space? How about other sections? Focus in on the snare drum sound, long a signature of recording engineers. Where does it fit in frequency and location? As we delve into specific mixing topics below, try to keep some of these big issues in mind.

Managing Levels. The most obvious function of a mixdown is adjusting the relative levels of your different tracks or MIDI parts. With few exceptions, the challenge is to strike a natural (or at least, effective) balance between the parts, so that no one instrument sounds overly pronounced, and no instrument is completely buried by the others.

In practice, however, there may be times when, for subtlety's sake, one or more parts are thrown "deep" into the mix; these parts may not be obvious on first listening, but they nonetheless contribute to the overall texture of the tune. And there are times—during a solo, for example—when you'll want to bring an instrument way out in front temporarily. We'll tackle these issues below, when we look at designing a soundstage.

The Virtues of Virtual Mixing. If you work strictly with a sound card's MIDI sounds—or with an external MIDI module or keyboard—your sequencer may be able to handle all of your level-adjustment needs. Almost all modern synthesizers change their volume in response to MIDI Controller 7 messages, which you can enter into a sequencer's event list window or often draw in with the mouse. (Multi-timbral synthesizers, which can play several different sounds at once, should offer separate MIDI volume control of each of the different parts.) As the sequence plays back, these messages will change the track levels in real time. Individual MIDI tracks can be faded in, raised up during a song's chorus, lowered during its bridge, and faded out at the end. You can re-record these levels over and over, and change them whenever you like; no matter what, all of your automated levels are retained in the program's memory, allowing you to build the mix track by track.

Many programs feature on-screen mixers that let you automate the playback levels of digital audio tracks you've recorded to your hard disk. The *pièce de resistance* is that you can also automate muting. With this feature, portions of tracks can be silenced, instantly and completely, for as long as you wish. (Muting is a perfect tool for cleaning up bits of unwanted audio, like lip-smacking on a vocal track or the hiss of a guitar amp heard before a song starts.)

Generally, sequencing programs let you use a mouse to adjust individual track levels, usually by dragging an on-screen fader knob, which is more intuitive than typing in Controller 7 values. But as the mouse can grab only one fader at a time, on-screen mixes are usually built one track at a time. Some programs offer a *grouping* feature, which assigns several faders to a group-master fader that controls the overall level for the group.

If you're one of those people who prefers to grab something physical, check out the JL Cooper FaderMaster or Peavey PC 1600. These external MIDI fader systems send out MIDI continuous controller data from each fader. Set each fader to transmit Controller 7 on a different MIDI channel, and you'll be able to adjust multiple tracks at once. High-end sequencers allow this incoming MIDI data to control their on-screen faders, providing a visual representation of levels. These fader boxes also sport buttons that can be programmed to control other functions, such as "record," "undo," or "mute."

Know When Less Is More. One of the toughest mixing challenges is to avoid pushing every track's level up to the max—that is, to know when less is more. This is especially hard if you're also the musician. After all, it's only natural that you'll want to promote each deft lick by pushing up the level. Unfortunately, the result can be a jumble of licks and hooks, all fighting each other.

Instead, practice the art of subtlety. Not every catchy riff needs to be pushed to the front; maybe your song would be better if the riff is kept quiet (or even eliminated) except during the bridge between choruses. In fact, seasoned producers know that a great hook should be used judiciously, so that the listener is *waiting* for it, instead of having it repeated *ad nauseum.* Conversely, many musicians tend (perhaps out of self-consciousness) to mix their own vocals too low.

Add Dynamics to Your Mix. Don't let the faders stand still during a mixdown. Pump the entrance of keyboards just a bit. Push the fader up at the end of each vocal phrase. Bring the delay in and out during the guitar solo. Make the background vocals swell when they hold a note. This kind of activity can give your mix a kind of life that most "perfect" recordings, with their compressed levels and precision performances, lack.

Equalization

An equalizer is a frequency-sculpting tool. Even a simple two-band EQ, such as the bass and treble section of a home stereo, can alter frequency response dramatically. For instance, boost the treble a little, and you can make music sound "crisper" and brighter; cut it a lot, and you can completely muffle the music. Cut the bass and music sounds thin, boost it a bit to add some "kick," or boost it to the max if you want sonic mud (and maybe a blown speaker). Between those extremes, boosting or cutting specific frequencies within a sound allows you to tame obnoxious resonances or bring out subtle character. For example, boosting a vocal in the 3 to 8kHz region can emphasize consonants, aiding intelligibility.

Project studios typically have a variety of different EQ resources. Many computer-based audio mixing programs offer *global* EQ that affects the entire mix, but this is not as useful as track-by-track EQ, since it's usually individual instrument or vocal parts that require EQ. More and more MIDI synths and sound cards offer built-in EQ, as do many external multieffects processors. With the latter, in particular, it's common to find relatively sophisticated types of EQ, such as three-band parametric. (Parametric EQ adjusts *which* frequencies are cut or boosted for each of the bands. On a home stereo, in contrast, the EQ frequencies are fixed—often affecting a wide band around 100Hz for the bass and around 10kHz for the treble. A *graphic EQ* has more and narrower frequency bands, but each band's frequency is still fixed.)

Equalization is a big topic—chapters have been written about it, and recording engineers regularly get into long debates about how EQ should be applied. But keep in mind these points:

- Start by making a "frequency inventory" of all the tracks you'll be mixing, to help you plan appropriate EQ settings. Strive for a balance of frequencies, which may require cutting as well as boosting different bands. Beware of *frequency masking*. An instrument that has been "EQ'd" to sound great on its own may not sound so great when it's in the mix. For instance, a muted guitar might sound nice and warm by itself, but become lost when played back with keyboards, vocals, and other tracks in the same frequency range. This is because our ears tend to blend simultaneous sounds into a single, composite sound. When several instruments emphasize similar frequencies, those frequencies accumulate—and can either become overbearing, or can cause one instrument to hide or "mask" another (see Figure 2).

- Excessive use of EQ is rarely a good thing. Too much low-frequency emphasis, for instance, can turn a mix to mush; too much high-frequency cranking can sizzle your listeners' eardrums. Most pros will tell you that less EQ is more, and that if you need to add a lot of it to get the sound you want, maybe you need to change the sound at its source. For instance, a "boomy"-sounding acoustic guitar, which

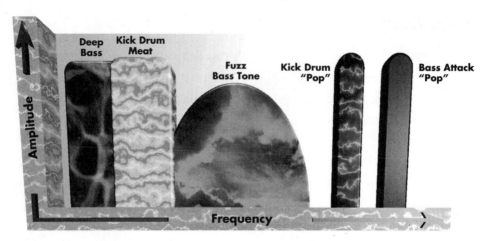

❷ *Figure 2 shows a rough graph of frequency vs. amplitude of the various voices used in a mix. This gives you a bird's eye view of the frequency spectrum and the potential for one element to mask another, helping you sort out what sounds you might add, subtract, or substitute.*

In the graph, there's no sign of a snare drum. That's because the snare drum, which contains a wide range of frequencies, would take up half the picture. If its frequency characteristics overlap those of the other sounds, why doesn't it drown them out? Because of its characteristics in the time domain: It decays to silence very quickly.

The meat of the snare drum falls right in the middle of the fuzz bass spectrum, but we hear the same fuzz bass both before the snare drum hits and after the snare drum decays. Because of its persistence, you hear the fuzz bass even though it's briefly obscured. Also, the drum's regular rhythm prompts the brain to expect to hear it at given intervals. The listener subconsciously gives less attention to the snare's spectral character than to its time-domain effect, namely the groovy beat.

might require you to cut a lot of bass, might be better fixed by adjusting the microphone placement to a less boomy position during the initial recording. If you're mixing MIDI tracks, try substituting different patches instead of using EQ.

• As we'll learn below, high-frequency response can also play a role, in terms of depth, for placing an instrument or voice on a soundstage.

Panning

Listen to almost any current pop or rock music mix and chances are you'll hear the lead vocal, bass guitar, and kick drum all panned dead center, so that they all project equally from your stereo's left and right speakers. ("Pan" is short for panorama, and refers to a sound's left-right placement in the stereo field.) If you have any difficulty telling what's coming from where, use headphones. You'll also hear, perhaps, an electric guitar panned somewhat to the left, a synth panned slightly to the right, and the rest of the drums (and maybe a piano) spread out across both the left and right panorama.

Panning MIDI instruments is quite straightforward. If you're using a synth-equipped sound card, an on-screen utility may let you adjust each instrument's relative pan position. Alternately, or if you're using one or more outboard synth modules or keyboards, you can turn to your MIDI sequencer. Just as MIDI Controller 7 controls volume, Controller 10 sets panning. Generally, entering a Controller 10 event with a value of 0 will move a sound to the left speaker and entering a value of 127 will move the sound all the way to the right. However, some older synths aren't set up to respond to MIDI pan information. In this case you may need to program the pan position for each part within the instrument (read the manual for more info).

When working with audio tracks, you'll either need to adjust pan on-screen (if you're using a multitrack hard disk recording system) or with an external mixing console. There's one big advantage to adjusting pan on-screen: You may be able to automate the panning, as described earlier for volume. This means that you can have individual tracks move left to right over time. (For an interesting effect, try bouncing percussion sounds between the speakers. It can really add life to a mix.)

Stereo Imaging

If you're working with stereo sources, such as a stereo keyboard or drum machine, you may be able to adjust their stereo *imaging,* not just their localization. If you own an external mixer, it's easy to hear the difference between imaging and localization.

Take a stereo source, such as a stereo synth string patch. Route the synth's left and right outputs to two separate mixer channels, and pan the channels hard left and hard right. Then play the sound, preferably from a sequencer so that your hands are free for mixing. Adjust both channels' faders for equally loud levels. Now pan both channels to the center; the synth is now localized in the soundstage's center and has the narrowest possible stereo image: mono.

Next, again crank the pan controls hard left and right. Since our source patch is stereo, you'll hear its widest stereo image. The maximum width is determined by the width of the source and/or of any stereo processing, such as reverb, that you're using. Now here's the payoff: You can adjust the source's localization without adversely affecting its image. Let's move localization to the right by lowering the left fader's level. Notice that as long as the left channel remains relatively audible, the wide stereo image is retained, even though the source's position has shifted to the right.

So in general:

- A stereo source's overall stereo image is determined by the degree to which the left and right channels are panned oppositely; the greater the difference, the greater the stereo width.

- To maintain stereo imaging, localize a stereo source by changing the fader balance.

- A mono source's localization is controlled by its associated pan pot.

Au Naturel or Au Bizarre? With some music, the aim is to localize and image instruments as they sound in real life. Most acoustic jazz and classical music engineers aim for recordings that reproduce live concerts as closely as possible. In fact, these engineers often record their ensembles live with just a pair of micro-phones, or a single stereo mic, which results in a stereo soundstage that closely matches the real thing. If natural is what you want, you'll probably want to use pan controls with discretion.

In contemporary production, however, panning is more a matter of feel than rule—who says you need a "real" perspective?

Designing Depth: Localizing Parts from Hither to Yon

If you own a DSP-equipped card, instrument, outboard processor or plug-in, chances are you've heard how reverb can make it sound as if you are playing at the bottom of the Grand Canyon, or in Westminster Cathedral. If they're offered, you've probably also tried different delay (echo) settings, along with chorusing, flanging, and other "swirling" delay effects. Maybe you've noticed how these types of DSP can add a professional sheen to tracks, especially vocals or other instruments that may have occasional intonation problems. Unfortunately, many home recordists end up using far too much of a good thing. In fact, more than any other *faux pas*, excessive reverb and delay are the number one giveaway of a home-brewed demo tape.

The challenge, therefore, is to learn how reverb and other processing can place instruments on your virtual soundstage. The depth positioning chart opposite lists a variety of ways in which you can localize an instrument near or far, using adjustable reverb and delay parameters (these are programmable settings that many multieffects processors, synths, and sound cards allow you to edit), as well as a few mixing, EQ, and microphone techniques.

Reverb Settings. Let's take a brief tour of what these parameters are, and how to apply them. Keep in mind that not every processor or instrument is going to offer every parameter, and that the names may vary (e.g., "damping" on one reverb proces-sor might be called "rolloff" on another). If you're in doubt, compare our descriptions with the ones in your product's manual.

Dry/Wet Level Balance. This sets the amount of the "dry" or unprocessed signal relative to the amount of reverb. A big "wash" of reverb will position a track toward the back of the soundstage—though it may make the track sound muddy. For an extremely far effect, go 100% wet. For up-front, in-your-face tracks, especially vocals, go 100% dry. For most applications, the setting will be somewhere in between (for example, 25% wet).

Reverb Decay Time. This parameter adjusts how long the reverberation lasts; a good way to hear reverb decay time is to trigger a single dry percussive sound, like a clave, and listen to the reverb. Decay time is often listed in seconds or milliseconds (thousandths of a second, abbreviated ms.). If two sources have similar dry and wet

DEPTH POSITIONING CHART

Parameter	To Move a Sound Back	To Bring a Sound Closer
Reverb Setting		
Dry/wet level balance	Increase wet level	Increase dry level
Reverb decay time	Increase	Decrease
Reverb algorithm/room size	Larger	Smaller
Reverb pre-delay	Decrease	Increase
Early reflection delay	Decrease	Increase
E.R. density & level	Increase	Decrease
Diffusion & density	Usually higher	Usually lower
Color/reverb EQ	Darker	Brighter
Delay Setting		
Delay	10 to >1,000ms, no modulation	None
Chorus/flange/phase shift	0.5 to 30ms delay + modulation	None
Mixing Functions		
Overall track level	Decrease	Increase
Hard left or right pan	Avoid	Okay
EQ Settings		
Presence (800Hz to 6kHz)	Cut	Boost
Microphone Technique		
Mic placement	Far-mic (add room mic)	Close-mic

levels, the source with the longer decay will sound farther back; if one source has considerably lower levels, then less decay will keep it positioned farther back. Keep in mind, as you crank the reverb past the five-second mark *'cuz it sounds so deep, man*, that many clubs have a natural reverberation of about a half-second; many of the best concert halls have a natural reverb of around 2½ seconds.

Reverb Algorithm/Room Size. Digital reverbs use different algorithms (mathematical models) to simulate arenas, halls, clubs, and other "rooms." Some reverbs also offer control over room dimensions (either cubic volume or length). Choose smaller algorithms or sizes to bring a particular track forward on your soundstage; choose larger ones to push them farther back.

Reverb Pre-Delay. In most live concert settings, the first sounds you hear follow a path directly from the instrument (or singer, or PA speaker, or whatever) to your ears. This is the initial dry signal, and is free of natural reverberation. Right on its heels, however, usually five to 75 milliseconds later, come a number of *reflections*, which are sounds that have taken a more circuitous route to your ears—bouncing off the stage, or a balcony, or a wall, before they arrive at your ears. The total of these reflections make up the overall reverberation.

To simulate the natural lag between the initial dry signal and the first reflections, many effects processors offer a *pre-delay* parameter. Closer objects have relatively long pre-delays between the time their direct sound reaches your ears and the time you hear any reflected signals. Consequently, increased pre-delay times can help position some tracks closer. (If large amounts of reverb or early reflections follow their respective delay times, however, a track may still sound far away, particularly if the overall mix is so complex that the pre-delay might not be noticed.)

Early Reflection Delay. Early reflections are the first reflections to reach your ears, and sometimes sound distinct, like little echoes. Like pre-delay, closer instruments will generally have a longer delay between the initial dry signal and the first early reflections.

Early Reflection Density & Level. Sometimes we never hear direct, dry signals from far-away sources. If someone's singing in another room or playing a tuba at the bottom of a canyon, everything we hear will be reflected. (Doesn't everyone hike with a tuba?) So, if a source must sound really far away, apply mega-amounts of early reflections and reverb, with no dry signal. Such distant sounds usually have very tight early reflection patterns, so increase ER density to move sounds backward; decrease the density to bring them forward.

Diffusion & Density. These parameters control the number of later reflections, the spacing between them, and how distinct they sound. For sweet, smooth reverb, engineers prefer high densities and high diffusion. Reducing them, however, can help position a track closer (though the reverb might sound more "chattery" than smooth). Typically, percussion uses more diffuse and dense settings than vocals.

Color/Reverb EQ. In real life, high frequencies tend to lose energy more quickly than low frequencies—that's why fog horns can be heard for miles. Reverberation from distant objects usually has fewer high frequencies, and sounds "darker" than reverb from closer objects. If your reverb offers a color or EQ control, try experimenting. (If it doesn't, and you're using an external processor, you could adjust the reverb's EQ at the mixing console.)

Delay Settings. Multieffects processors usually offer a delay (or echo) function. Delay times greater than 50ms will be heard as distinct echoes, which, when mixed with reverb, simulate super-distant placements.

Chorus, flange, and phase shift are all *modulated* digital delay effects, which means the delay time varies in a (usually) slow, periodic manner (in some devices, the delay time varies randomly, or in accordance with input level changes). The swirling sounds they create can sometimes make the sound appear wider than, or emanating from in front of, the speakers.

Mixing Functions. Mixing a track dramatically quieter or louder than other tracks can position it without having to use reverb or delay. As long as there's enough musical "space" around the track—that is, there aren't so many similar sounds playing that the track gets lost—it can be placed way back on the soundstage and still sound defined.

Panning can also influence depth. In real life, you'll hear a source in just one ear *only* if it's right up against your ear. To position a mono track far back in the soundstage, don't hard-pan, regardless of how much reverb you add to it. (For a stereo source, like a synth or drum machine, it's okay to pan *both* outputs left and right respectively since the sounds have already been panned to intermediate locations within the machine.)

EQ Settings. Intelligibility is a hallmark of up-close instruments or vocals. Consequently, boosting a track's "presence" helps bring the track forward. (Presence correlates to the range of frequencies from roughly 800 to 6,000Hz; this is usually adjustable with either a midrange or high-frequency EQ control.) Cutting the same frequency range adds authenticity to tracks already positioned by reverb or room ambience.

Microphone Placement. To add natural ambience to audio tracks that have been recorded with a microphone, place the mic at a distance of three feet or greater from the source. You can also mix in the sound of a "room" mic—a microphone six feet or farther from the source—to position a source farther back. Keep in mind, however, that it's impossible to remove room sound once it's been recorded with a track, which is why it's often better to add the reverb artificially during mixdown.

Preplanning Your Mix

With so many ways to localize tracks on your imaginary soundstage, it's easy to wind up with a cluttered mix. It's also easy to position tracks formulaically, without thinking. Fortunately, some preplanning can open new creative doors onto your soundstages. Preplanning your soundstage is cheap, and makes good engineering sense. After all, you wouldn't build a *real* stage without blueprints, would you?

Sketch Placements. Many engineers start off a mix by panning tracks in a familiar manner—for instance, kick, snare, bass, and lead vocal center; keys to the right; acoustic guitar to the left; etc. Feeling inspired, they may vary one or two of these placements, but not much. Similarly, they'll use reverb with a familiar approach: wet-sounding snare, a moderate amount of reverb on the vocals, no reverb on the bass, and so forth.

If this sounds familiar, you, too, may be in need of a creative kick. So next time you mix, reach for a pencil before you reach for a pan control. Make a diagram of the

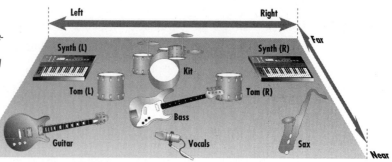

❸ Sketch out your instrumental placements before you start to mix. Not
only will this help you identify potential
problems, but you might also find
that using a visual approach gives
added inspiration.

soundstage you are about to create (see Figure 3) and sketch out the following details
for each track:

- Left-to-right placement (pan)

- Depth positioning (reverb amount and other positioning techniques)

- Stereo width (for stereo sources, or tracks with stereo reverb)

- Acoustic space (reverb room algorithm or size)

By using your mind before you use your ears, you'll approach the mix from a different point of view. As an added benefit, the visual reference map can help you
avoid a cluttered mix.

Plan Acoustic Spaces. Remember, the primary way to control depth positioning is
to adjust the amount of reverb. If your aim is a natural-sounding mix, say for acoustic
jazz or classical music, usually it's best to use just a "global" reverb setting—with a single room algorithm (hall, cathedral, etc.) and a single decay time that's applied to all
instruments. Then, the simplest way to position tracks near or far is to adjust each
track's reverb send level: more reverb to position a track to the rear, less to bring it forward. After adjusting the reverb sends of the individual instruments, pull back on the
master reverb output level until you can hear all of the instruments clearly.

However, if you're mixing modern pop/rock music, and your sound card or synth
allows for effects processing on a track-by-track basis (or you're using a mixer in conjunction with multiple processors), it's common to combine a variety of room
algorithms and decay times. "Big" drums, for instance, typically call for an arena algorithm with a long decay time; an up-front vocal might call for a plate algorithm with a
rapid decay. You can also achieve "big" sounds using a smaller room with a long
decay, and setting the wet/dry balance to relatively wet. (However, if you're unlucky
enough to have a cheap, gritty-sounding reverb, like the ones built into many sound
cards, it's advisable to keep things on the dry side.)

When using multiple room algorithms, it's helpful to draw size diagrams for each
track in your soundstage, as shown in Figure 4. Doing so may inspire novel-sounding
mixes as well as help you avoid clutter. And it will make it easier to identify tracks
that can share reverb channels—an important consideration for anyone with fewer
reverb units than tracks.

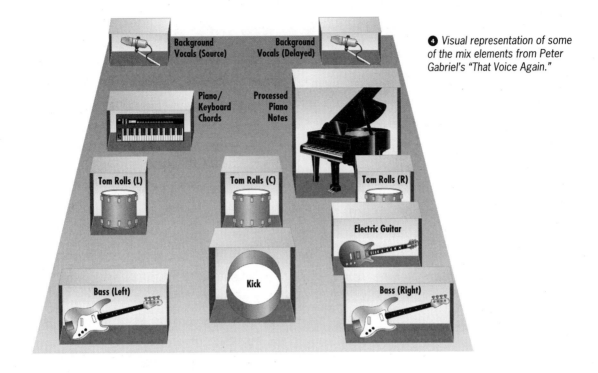

❹ *Visual representation of some of the mix elements from Peter Gabriel's "That Voice Again."*

Non-automated Mixing. Automated mixing is wonderful, but some of us are still doing it the old-fashioned way, by moving the mixer's faders up and down during playback (see "Digital Multitrack Automation" for automating with digital recorders). Try keeping track of levels with colored pencils. If there's a smooth surface, you can mark directly on your mixer's panel with a grease pencil. (Make a test mark first in an inconspicuous corner and try rubbing it off. If you use isopropyl alcohol to clean the marks off, also test it first. Some chemical cleaners will degrade some types of plastic panel.) Or put white artist's tape alongside the faders, and mark on the tape.

Plan your mixing moves. Try recording the tracks with natural dynamics, so you don't have to change levels constantly. If you're recording a quiet sound for the background, record it at a lower level, consistent with staying above the noise floor. Try to record tracks so that there are gaps between the events so you can make your moves without panicking.

Quality Counts. Ultimately it's the quality of your music that really counts, and not so much the mix. Listen long enough to any radio station, and you'll hear mediocre songs packaged oh-so-nicely in gloriously slick mixes. While a great mix might make a hit, a great mix does not make a great song.

But the flip side also applies. It's not just enough to have a great song, especially if you're hoping to have it noticed. Record label execs frequently pass up absolutely terrific material because of a poor mix, even though they're expected to listen to the music and not the production. So if you believe in your music, think as a playwright might, and choose to have your story told on the most appropriate and flattering stage you can create.

Digital Multitrack Automation

How many times have you been working on the perfect mix, only to blow it in the last few seconds by forgetting to bring a particular instrument down (or up)? Those with big-bucks automation systems or automation software can flip to the next section. For the rest of us—well, don't feel bad: It's a common problem. One of the solutions is automated, "snapshot"-style mixing.

A mixer that performs snapshot automation stores all control settings (or at least the level settings) as a recallable snapshot. As the song plays back, each time the mix needs to change you move the controls to their new positions and store the settings as a snapshot (this approach assumes that your mixer settings are relatively constant for several measures at a time). You can then manually superimpose the occasional dynamic fader change (such as ducking a bad note, turning up an effects send briefly, etc.) without having to think about the rest of the control settings.

Some mixers use manual snapshot recall from a front-panel keypad. Others are triggered by MIDI program changes, and still others use an internal sequencer, usually synchronized to MIDI or SMPTE time code, to trigger a series of snapshots in order. In any event, the end result is a more precise and recallable mix.

However, mixers with this capability are relatively expensive. Even low-cost outboard boxes (e.g., the Niche ACM) might be too much for a budget already strained by buying a hard disk or digital tape recording system.

Fortunately, there is a low-cost snapshot solution using gear you may already own. This method works with multitrack recorders that use digital tape or hard disk technology.

Digital Multitrack Snapshots. One side effect of digital recorders (tape or hard disk) is that they have brought back the art of premixing, or bouncing tracks. Part of this is from necessity; because track counts are typically limited, you'll probably need to premix if you want a complex, layered sound. But also, digital audio makes quality premixing possible. Unlike premixing with analog machines, no longer does the process build up noise and distortion.

Our snapshot method uses two unused tracks on your digital tape or hard disk multitrack to hold the final mix. The downside is that with an eight-track, you now have only six tracks available for basic recording, so this method works better if you have 16 or more tracks available. However, the advantages of this type of mixing are quite compelling (especially with 16 or more tracks), as it gives you snapshot mixing "for free." Here's how to do it:

1. Connect the system as shown in Figure 5, where the recorded tracks go through the mixer to two unused tracks.

2. Go to the beginning of the song, put the unused tracks into record, and start mixing.

3. As long as the mix is okay, keep recording.

4. As soon as you want to change the mix, stop the multitrack and rewind prior to where you wanted the mix to change.

5. Play back the multitrack and adjust the mix levels, EQ, etc.

6. Rewind the multitrack to about 10 seconds before the point where you want these changes to kick in.

7. Punch in on the mixdown tracks at the precise moment where the new mix should take over. Unlike analog tape, digital tape doesn't leave a punch-out gap to fret about. (You can optionally use a rehearse mode, if available, to set and audition the punch points.)

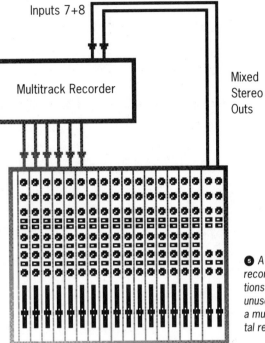

Inputs 7+8

Multitrack Recorder

Mixed Stereo Outs

5 *A mix can be recorded in sections onto two unused tracks of a multitrack digital recorder.*

8. Repeat steps 4 through 7 until the entire song is mixed.

9. Your final mix is now located in what had been two unused tracks. Bounce them over to DAT (or whatever your mastering machine might be) using the analog or digital outputs, and you're done.

This is really a great way to mix, as it gives the advantage of intuitive, real-time mixing up to the point where you need to make a change, at which time you reset a few controls and carry on.

Sure, fully automated mixing that re-creates all fader and control moves is convenient and effective. But when you have to do more with less, the above approach can give you a perfect mix without too much effort—and with very little cost.

Mixing with Effects

Synthesizers, audio recorders, and effects processors are turning increasingly digital, but when it comes to hooking all this gear together, it's frequently still an analog mixer that does the job. The flexibility, number of patch points, and hands-on control remain appealing, as does the cost. However, in many ways these venerable boxes remain underutilized, particularly regarding effects. Here are several tips designed to increase the synergy between mixing and signal processing.

Full-Feature Effects Returns. If you have enough mixer channels, bring your reverb returns back into two mixer channels rather than the dedicated effects returns. Using mixer channels gives you more control over the returned signal (EQ, panning, sends to other effects, etc.) that the returns usually don't have. When using this patch, be careful not to turn the return channels' corresponding send(s) up, as this would create a feedback loop.

Keeping Your Straight Sound Straight. Although mixer input channel effects loops are convenient for patching in effects, some processors alter the straight (dry) sound as the signal passes through them. Fortunately, you can preserve the integrity of the straight signal yet still add the desired effect; the patching option you'll use depends on your mixer setup.

In any case, the first thing you need to do is derive a send from the input channel signal. There are three ways to do this:

• If your mixer has separate loop send and receive jacks for each channel, patch into the send. This should not break the normalled channel connection that allows the straight signal to pass through to the stereo mixdown bus.

• Newer mixers often use a TRS (tip/ring/sleeve) stereo jack to handle the effects send/return. Plugging a cord halfway into the jack (so that the plug tip contacts the jack tip) should provide a send without breaking the normalled connection.

• If you don't have loop jacks, then use an aux bus to provide a send from the channel.

Patch the send into the processor, which should be set for effected sound only (no dry signal). Then bring the effects output back to a separate channel, or to two channels if it's a stereo effect, and mix in the desired effects blend at the mixer (dry sound on the original channel, processed signal on the additional channel or channels). As a bonus, using this approach lets you modify the effects signal with panning, reverb, aux sends to other effects, and all the other input channel options.

Better Reverb Pre-Delay. For more control over your reverb sound, patch a delay line between the effects send and reverb input. This provides more control than the pre-delay found in most reverbs. For example, adding a bit of feedback to the delay line can create a more complex reverb effect. Again, set both the delay and the reverb to an effect-only (wet) setting.

Bigger Stereo Images for Piano and Guitar. Here's an effect that Alex de Grassi uses a lot on his guitars to create a wider stereo image with two mics, but it also works well with piano. Split the main right and left channels to two additional channels, using either a Y cord, a direct out, or an effects loop send. Pan the main left and right channels hard left and right, and center the other two channels (see Figure 6) but bring their levels down about 5–6dB (or to taste). This fills in the center hole that normally occurs when the two main signals are panned to the extreme left and right.

While you're at it, experiment with adding reverb in different ways—only the main channels, only the middle channels, weighted toward the left or the right, etc.

Save an Effects Send. Most stereo reverbs are not true stereo, but sum the two inputs together to mono and synthesize a stereo field from that. Therefore, there's no real need to use two sends, as the original stereo imaging is lost anyway. However, do use stereo returns.

Undead Compression. If you're using compression as an effect (as opposed to preventing tape saturation or some other "utilitarian" application) but don't like that squashed sound, patch the signal to be compressed through the main

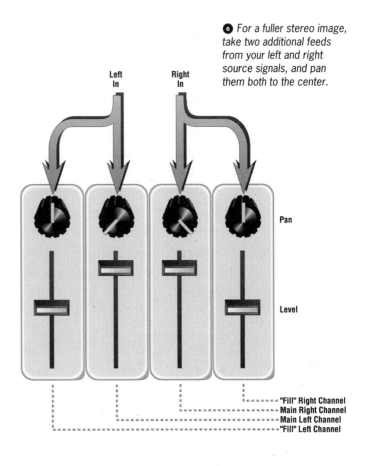

❻ *For a fuller stereo image, take two additional feeds from your left and right source signals, and pan them both to the center.*

channel and send its direct out or effects loop out to the compressor, then return the compressor to a separate channel. Use the compressed channel as your main signal, then bring up the unprocessed channel to restore some of the dynamics.

Aux Bus Fun and Games. Get creative with your aux send processing—no law says you can use only reverb. Some of my favorites:

- Add a mild distortion device (tube preamp, etc.) and send it drum tracks, bass, whatever. The distortion can add a nice edge and warmth; bring it back at a fairly low level just to add a bit of "crunch." Or patch the send to a guitar amp, mic the amp and the room, and bring the mic signal back into the board as an "effects return."

- Feed one aux bus to a vocoder carrier input, and another aux bus to the vocoder's modulation input. Note that this requires a real vocoder with two inputs, not the digital simulations that have only one input (with perhaps an additional input for MIDI control). This allows any signal to modulate any other signal, which can provide a very cool effect if you use something percussive for the modulation input and use it to trigger a more sustained part, such as bass or long piano chords, which feed the vocoder carrier input.

- Miss that Roland Space Echo tape delay you sold at a garage sale years ago? For that unmistakable tape echo sound, patch a pair of aux sends to an analog, two-track, three-head recorder set for record. Roll the tape, and patch the playback head outputs to the effects returns. You'll end up with a slapback echo that has that warm analog quality. If the deck has multiple speeds and variable pitch control, so much the better. For a truly grungy delay, use this technique with a three-head cassette deck.

- For some audio excitement, I often patch an "audio exciter" into an aux bus send/return rather than use it inline on the entire mix. Although these types of devices were originally intended to process an entire mix, these days it's not uncommon for samples to already be "excited" (particularly drum samples), and adding more exciter on top of that can give a really tinny sound. Adding an exciter selectively using the aux bus controls gives a much more controlled mix. Note: Some of these devices do not allow for cutting out the straight signal so that you have processed sound only—a necessity when using an exciter in an aux loop.

MONITORING TIPS

What's the biggest impediment to a well-balanced soundstage? Probably the loud-speakers you use as you record and mix, and the room in which you work. Almost all speakers accentuate certain frequencies and underrepresent others. Even with expensive "studio monitor" speakers, the acoustics of the listening room conspire against them—especially with the proverbial bedroom/spare room/living room/garage studio that lacks acoustical treatment. For these reasons, the monitor system is universally the least accurate component of any studio.

The problem gets even worse if you use a typical pair of home stereo speakers as monitors. Maybe you think they're fine, because all your commercial audio discs sound great when played back through them. Unfortunately, that's the problem: Most home stereo speakers are designed to sound good, not accurate, so they'll frequently boost the low end for a bit of thump, and boost the high end for a bit of sizzle. Mixes made on speakers like these can sound bass-shy and muffled when played back on somewhat more accurate speakers.

The good news is that there are a number of ways you can compensate for your speakers' shortcomings, to help ensure a mix that sounds good in lots of different playback settings.

Use a variety of references. Listen to your mix through a car stereo (the favorite reference place for many producers). Car stereos have funky frequency weightings to compensate for the acoustics of the car interior. This is a good test to see just how flat your mix really is—if nothing totally unexpected (typically a low-midrange guitar) suddenly dominates a mix that sounded relatively subdued in the studio, you're in the clear. In addition, the auto environment is rife with background noise. Songs with a wide dynamic range can fall beneath the noise floor during the quiet parts. This isn't to say you should emulate the hyped-up, piped-in, compressed-tokingdom-come sound of your local hard rock radio station. It just drops a clue that the quiet parts may need something else to help them pierce through a noisy environment, such as a boost to the high frequencies or less reverb.

Play your mix on a boom box, a friend's hi-fi, and through a WalkThing. Let your alternate references teach you how to mix. If the music sounds good in your studio but too bassy on your friends' stereos, you may be overcompensating for your monitors' lack of bottom end, and need to learn to mix with less low end. Also, home stereos—especially those with built-in effects or surround sound—have a way of exaggerating audio spatialization. That dreamy echoed guitar solo can turn into a splatter of disjointed frequencies on a loud home stereo, indicating that you should go back and pull back on the effects.

If your mixes sound excessively bright everywhere but in your studio, perhaps your monitors' tweeter controls are turned way down, or worse yet, the tweeters are fried. If your music sounds good through most or all of these references, you know you're on the right track.

Use professional near-field monitors. For $200 to $500, you can get a pair of monitor speakers that are more accurate than many expensive home stereo speakers. Be sure to buy near-field monitors, designed to be placed within a meter or two from where you sit (to minimize interference from the room's acoustics). If you're serious about your music, good monitors are money well-spent.

When you're seated at your mixer, the space between the two speakers and your ears should form an equilateral triangle, about one meter on each side. Angle the speakers slightly toward your ears, and keep them at eye level.

Compensate for your studio's compromises. Since you probably don't have several thousand dollars to have your studio professionally "tuned," when using near-field monitors the best way to reduce the impact of your room's acoustics is to hang drapes or acoustical foam (like Sonex) over hard walls and windows, throw carpets on hard floors, and generally do your best to "deaden" the studio.

Get intimate with your tracks. Hear beyond the obvious. To ensure that the reverb on your lead vocal is decaying smoothly, listen just to that track (all other tracks should be muted). Solo all of your MIDI tracks one by one, and mute, erase, or repair any musical clams. If you're recording audio to hard disk or tape, clean up the tracks before you mix, by erasing unwanted sections, or if necessary by using a noise gate, which can silence hum, clicks, and other low-level junk. Many on-screen audio editing programs offer built-in gating; several compressor/limiters designed for use with a mixer also include gating functions.

What about headphones? Don't monitor exclusively through headphones—late at night, maybe, or when you're recording with a microphone. But committing yourself to a mix while monitoring through headphones is a mistake. On the other hand, you can hear details through headphones that you might miss through speakers, so they're a good way to "proof" individual tracks and complete mixes for stereo imaging, any glitches or pops, and other details. For best results, go back and forth among several different sets of speakers as well as headphones to get as complete a picture of your mix as possible.

(continued on next page)

Vary the perspective. Listen at moderate to quiet levels. (It's okay to listen at loud levels for short periods, but be aware of room interference and your ears' health.) To avoid "listener fatigue," take a 10-minute break every hour. Don't make final decisions at 3:00 A.M. after hours of mixing. Close your eyes. A common engineer's trick is to go behind the speakers and put your head in between them; this can reveal unheard details. Walk around the room. Turn your back to the speakers and listen carefully. Any of this might let you hear something new.

Spare the loudness. While a professional power amp is best, a home stereo amp or receiver can be used for near-field monitoring—assuming it's a quality product with 30 to 50 watts per channel minimum. If you go this route, beware the temptations of the Loudness button. This is actually a variable EQ circuit, which boosts highs and lows at quiet to moderate listening levels, and it can make you think you're mixing with more bass and highs than you have. Make sure the loudness control is off before doing any critical mixing.

Burning

6

Now that CD recordable drives and blank CD-R discs are affordable to even those on limited budgets, what computer-owning musician can resist? Demos recorded on CD have the same pristine quality as a DAT tape, but the person you send them to won't need anything other than a standard CD player. If you have a few fans who'd like to buy your album, but not enough to justify duplicating CDs in mass quantities, with a CD-R you can "burn" (record) just enough discs to satisfy demand. Recordings archived to CD-R won't degrade as quickly as analog or DAT tape (though exactly how long they *will* last is a matter of some contention).

Most CD-R drives (also called CD writers) can produce not just audio CDs but CD-ROMs that hold up to 650MB of data, so you can use them to archive completed digital audio projects, create sample and patch library discs, or even run off some discs for sale—there's probably somebody out there who'd pay big bucks for that 300MB of didgeridoo samples you've spent years collecting and indexing. You can even back up your hard drive to CD-R as well, now that CD-rewritable discs (which can be used over and over, as opposed to being "write once") are available. A CD-R drive can also hold multimedia projects such as enhanced CDs, karaoke, and video CDs.

Equipment You'll Need. You don't need a lot of fancy equipment to record your own CD (see Figure 1). But you do have to make sure the gear you have is optimized for the process of creating CDs.

Burning CDs requires a powerful computer—many a disc has been wasted because a system was too slow overall, or had a weak link in its speed chain. On the PC side, you'll need:

❶ *You don't need $50,000 of equipment and a fancy studio to make a CD at home. This is the desktop where author/musician Dave O'Neal composed, recorded, mixed, and mastered his CD.*

- A CD-R drive supported by the software you plan to use

- A SCSI or PCI hard-disk controller card (most newer designs place these controllers directly on the motherboard)

- A 100MHz Pentium or faster CPU

- At least a 1.5GB AV-compatible hard disk (audio/video drives do not pause for thermal recalibration in the middle of recording a CD so that data is transferred at high speeds, without interruption)

- A minimum of 32MB of RAM

- CD-R software (usually included with the recorder)

- Blank discs for recording

On the Mac side, in addition to the big hard drive, CD-R drive, CD-R software, blank discs, and RAM, you'll need a Centris, Quadra, or Power Mac operating under system 7.5.1 or later. That's the big picture; let's look at some details.

CD-R Drive. Connecting a drive is usually just a matter of plugging it in to the appropriate computer port. If you buy a writer designed for internal mounting, you'll have to open your computer, but a SCSI conflict is the most difficult problem you're likely to encounter. Usually, fixing this is as simple as changing an ID selection switch to a different ID number. In extreme cases you may have to open the unit and move jumpers. (See sidebar, "Unraveling SCSI.")

Mount a CD-R unit on a stable surface, as close to level as possible. Nearly all manufacturers provide a list of do's and don'ts (usually printed on a separate slip of paper in the box or in a help file), as well as lists of compatible SCSI controllers, drivers, and media. Most CD-Rs use a disc caddie rather than a retractable tray. The caddie keeps discs a bit more stable, which is crucial for error-free writing.

SCSI Controller. All Macintoshes from the Mac Plus on have included SCSI, so the following pertains only to Windows systems.

Even if you already have a SCSI card, you may need an extra one if you've maxed out your current SCSI chain with a gaggle of hard drives, scanners, removable drives, and the like. If you have an old or cheap controller (like the one that came free with your scanner), you might want something better—an inexpensive card *might* work, but you're probably better off buying a good Fast or Ultra Fast SCSI-2 controller by Adaptec, Future Domain, or Atto. It will handle your CD-R and provide improved SCSI-2 hard disk performance. To all you IDE/ATA folks out there, sorry, but you'll have to move up to SCSI to burn your own CDs.

As for cables, SCSI cables are usually pretty short (two or three feet) but you can get them up to six meters (19.6 feet), which is the limit on the total cable length in the SCSI chain. For best results, put the CD-R next to or inside the computer. And don't forget about proper termination (see sidebar, "Unraveling SCSI").

Hard Drive. You'll need a hard drive with less than 18ms average access time, sustained data transfer rates above 800KB/sec (for double- or quad-speed recording, you might want something a bit faster), and a high rotational speed (7200 RPM or higher is optimum, although somewhat noisier on average than drives with slower rotational speeds).

Depending on the software and type of writing you're doing, you may also need enough free hard disk space to hold a complete one-for-one copy of the data you intend to write, plus a little extra. So to fill a 650MB disc, you need roughly 1,300MB free on your hard disk drive. Since it's important to leave some free space on your drive, 1.5GB is a good choice for a minimal system.

Software. CD-R software comes in flavors for just about every type of computer and level of experience. Packages are available for Mac, Windows 3.1/95/NT, OS/2, Novell, Unix, Sun-OS, and DOS. Most manufacturers sell separate versions for different platforms.

Just about all CD-Rs bundle software that lets you pick which files to put on disc, does some special formatting, and controls the recorder's mechanical operations (with the exception of *packet-writing* software, you can't just copy files to a CD-R as you can with other media). Most mass duplicators require a *disc-at-once* master, where the entire CD is written at once, as opposed to a *multisession* disc, where different sections of the disc are written on at different times.

Burning Basics. Assuming your hardware meets the needed requirements, using CD-R software is pretty straightforward. In most cases, you simply tell the software what type of disc you want to make (Red Book audio, CD-ROM, multimedia, etc.), drag and drop filenames or complete directories into a special window using a file-manager-like utility (many PC packages actually use Windows' File Manager itself), specify a writing speed (1X, 2X, 4X, etc.; higher speeds place higher demands on the hardware), and then tell it to go. You can usually initiate a "test run" first, where the system determines whether the hardware is up to the task at the specified write speed. If your system passes the test, odds are that the CD-burning process will create something other than a coaster. Caution, though; these tests are not infallible, and any one of a number of hardware ills—from something as obvious as a power outage to as subtle as a misplaced interrupt—can interfere with the writing process.

At 1X speed, it takes a little more than "real time" to write a CD (e.g., if you're recording 30 minutes of music, the process will take a little over 30 minutes). The extra time is required for the CD writing process to close out the file after writing. A 2X writer takes about half the time, a 4X writer ¼ of the time, etc.

Of course, a lot more is happening under the hood, but the software should make this all pretty transparent. If you're curious about the CD-burning process, refer to the "For Further Reading" section at the back of this book.

Buffer Underruns. When writing, CD-R drives need a continuous stream of data. An interruption that lasts long enough for the cache to run out of data, called a "buffer underrun," ruins the disc. To reduce the likelihood of buffer underruns, observe the following:

- To handle momentary breaks in the data flow, CD-Rs have onboard RAM caches of 512K or more. Buffers of at least 1MB are preferred for CD recording.

- Thermal recalibration of your computer's hard drive is the most common cause of buffer underruns. This short pause occurs while the drive realigns itself to correct for minor variations caused by temperature fluctuations. To avoid problems with writing CD-Rs or performing other real-time digital audio operations, buy an AV

UNRAVELING SCSI

If you bought your PC at the local computer mart, odds are your hard disk drives are IDE (Integrated Drive Electronics), EIDE (Enhanced version of the same), or ATA (AT Attachment—same thing as IDE). That is, they're not SCSI (Small Computers System Interface).

Unfortunately, in general IDE just doesn't cut it for CD-R and digital media. IDE drives and controllers are less expensive and therefore more widely available. But they aren't very fast, with data transfer rates in the area of 2MB/sec. Every CD-writer available today is a SCSI device.

The original SCSI-1 spec isn't necessarily bad, it just isn't great. CD-R units that write at 2X, 4X, 6X, and 8X speeds pretty much require a Fast SCSI-2 connection (roughly 10MB/sec data throughput) and an AV drive.

With Windows, installing a SCSI controller board is as simple as installing any new card—just plug it into an empty slot. From then on, however, things can get tricky. If you're connecting internal drives, use the ribbon cable that's usually supplied with the drive. External drives connect using thick cables that can't exceed 19 feet in length and usually come in one-meter lengths.

SCSI is a daisy chain in which each device connected to the bus, including the host ID card, has its own unique ID (with several cards, each card also has its own host number). These numbers usually are set with DIP switches, jumpers, or, if you're lucky, click-switches that cycle through the possible numbers. Having unique IDs makes it easy to add or remove devices from the chain, since you don't need to configure anything as long as no two devices use the same ID. If your computer has a SCSI controller on the motherboard, it's probably set as host 0, device 7; your internal SCSI hard drives probably are numbered sequentially starting with 0. Unfortunately, this setup can cause problems with certain combinations of CD-R writers and software.

When you power up, the SCSI controller polls the system by checking ID numbers to see what's connected to the bus. If two devices use the same number, most often nothing works. If one of the devices is a bit slow on the uptake, the system might boot anyway because the slow device failed to respond to the controller in time. Afterward, though, you can be sure the ID conflict will cause trouble.

Assuming all the devices have a unique ID, the controller then assigns drive letters to the devices, which is why drive letters may change when you add new devices to the chain. With Windows, you might need to edit your config.sys file by changing the last drive line to LAST DRIVE = X if you get a NOT ENOUGH DRIVE LETTERS error message during boot-up.

SCSI behavior can sometimes confuse CD-R software. For example, some programs can build and test a virtual disc that contains only a list of file locations rather than copies of the files themselves. If you attach a new device to the SCSI chain, the drive letters will shift so that drive D becomes E, E becomes F, and so forth. Now the virtual list points to drive letters that no longer match the drives on which the files reside.

Termination, which reduces reflections along the transmission line that can confuse the system, is yet another SCSI quirk. Theoretically, the first and last devices in a chain must be terminated or the whole thing won't work. However, if you read the fine print, you'll notice that it says the chain "won't work properly," which is slightly different.

Some devices are self-terminating, some must have a separate terminator if they're last in the chain, and some—like some Macintosh computers—don't seem to care (provided the cable length between devices is 18 inches or less). In a chain that consists of only internal or external devices, the host adapter usually can handle its own termination. But if you have both internal and external devices, the host adapter is considered to be in the middle of the chain and the devices on either end must be terminated. The longer the cables (and the more varieties of cables and devices) in the chain, the more important proper termination becomes.

Three kinds of terminators exist: passive, active, and differential. Buy an active terminator and stick it on the last external device—usually the CD-R writer. If that doesn't make the SCSI chain behave, try taking the terminator off, try swapping cables, and try shorter cables.

Once the SCSI board is set up, you'll need to install the proper software. Since, from Windows' point of view, a SCSI card is yet another alien device, you need an ASPI (Advanced SCSI Programming Interface) device driver. This software usually comes with the SCSI card, but some companies sell ASPI drivers separately. Both Adaptec and Corel offer excellent SCSI controller software. (Most CD-R software and hardware works with either brand and, in some cases, actually requires one or the other.) Micro Design also has SCSI controller software that works with their writer/hard disk combination.

Unfortunately, this is another one of those areas where the process either goes smoothly or leaves you pulling your hair out in clumps. Many PCs come from the factory with built-in SCSI controllers and software. This frees up a slot, but it can cause problems if the CD-Rs depend on the quality of the controller, the system configuration, and the software used. Sometimes the software doesn't recognize the controller or the particular ASPI driver. Sometimes the system insists on checking the built-in controller first, and adding a second SCSI board just compounds the problem.

Be patient and read all instructions carefully. With luck, you'll plug in the board, install the ASPI drivers, and the system will work like a charm. On the other hand, you might spend an afternoon changing SCSI IDs, installing and re-installing software, and swearing. (Be prepared for the slight possibility that your particular configuration simply can't be made to work.) Once you have SCSI figured out, you'll never want to go back to IDE or anything else. If you already have SCSI drives and have mastered the art of attaching new devices, then you're likely to find hooking up a CD-R as easy as plugging in the box.

hard drive, which detects continuous data transfers and postpones thermal recalibration until it's finished. Currently, all AV drives are SCSI; PC owners who want to use EIDE drives have to buy the fastest they can get and keep their fingers crossed.

- Reduce the number of devices on the SCSI chain to a minimum, and make sure the bus is properly terminated.

- Avoid running other applications while burning CDs, since every application that accesses the disk narrows the bandwidth available for transferring data to the CD-R drive. Even disk cache utilities that would speed up most applications can screw up a CD-R burn, because you want data to flow in a continuous stream, and not be cached on the way to the CD-R. Keeping Mac control panels and system extensions to a minimum will also help (use Extensions Manager to disable all extensions that aren't absolutely necessary). For Windows machines, disable add-ons like Norton Desktop or Microsoft Office Fast Find (which constantly "polls" your hard drive to maintain a current catalog of files), and also turn off CD-ROM Auto Insert Notification. To do this, open the "System" icon in the Control Panel and select "Device Manager." For each item under CD-ROM, select the device, click on the "Settings" tab, and uncheck the "Auto Insert Notification" checkbox.

- If the CD-R burning software allows, increase the buffer size to reduce the chance of an underrun (this buffer is not the same as the computer's disk cache).

- Turn off any screen savers. These can cause havoc with some digital audio programs. If you're going to be away from the computer and don't want to turn off the monitor, just reduce the brightness.

- Don't do anything else with the computer while recording. With Windows machines, there may be some programs running in the background, such as routines that monitor software installations. Press CTRL-ALT-Delete to bring up a list of currently active programs. Click on any program you don't need and select End Task.

- Defragment your hard drive, especially with on-the-fly recording. (With "on-the-fly" recording, the CD-R picks up data from the hard drive as needed in the process of creating the CD. With "disk image-based" recording (see below), the program creates a file that duplicates the data exactly as it will go on the CD, thereby alleviating some processing that can be problematic with slower computers.) Even with a fast AV drive, you should defragment perodically. Fragmented drives slow disk access speeds and eventually lead to buffer underruns. Occasionally reformat the drive you use for CD images to guarantee a clean slate.

- Record from an ISO "disk image" file rather than on-the-fly. Running a disk image will often let you know if an underrun would've occurred under real-time burning conditions, thereby avoiding the creation of a coaster. Exact disk images also are great for archiving projects that will be used to create multiple discs. The potential downside is that you could need as much as 650MB for a CD-ROM or 740MB for an audio disc when using a 74-minute blank.

- Always scan for viruses before burning CD-ROM discs. Commercial software has been released with viruses immortalized in polycarbonate. With a little extra care, you can avoid doing the same. Once you know a drive is virus-free, quit the virus scanning program so it doesn't unexpectedly "wake up" while recording a CD.

- Watch for virtual memory settings that cause swapping, unusual network activity, and background data downloads or faxes. Because Windows 95 is a multitasking operating system, it makes extensive use of *caching.* Caching shuttles data into memory, where it hangs out until the cache (also called buffer) fills up, whereupon the data's "released" (e.g., to the hard disk or CD-R). Caching interferes with sending data to a hard disk or CD-R as a continuous, unbroken stream. As a result, most digital audio programs recommend limiting the cache memory. Furthermore, Windows 95 thinks it's doing you a favor by dynamically resizing the cache, but the resizing process takes additional time and can cause more problems.

You can fix this by modifying the SYSTEM.INI file (located in the Windows folder), but first, back it up to a floppy disk. (If you see more than one file called System, right-click on each one and select Properties to see the full title. You want System.ini, not System.dat.)

To change the virtual cache size to a fixed size, go to Start > Run. Type Sysedit at the prompt, then click on OK. You'll now see all the system configuration files. Click on SYSTEM.INI so it becomes the active window. Find the [vcache] section (or create if necessary) then add, or edit, the following lines:

[vcache]

MinFileCache=2048

MaxFileCache=2048

This limits the cache to 2MB. With 32MB minimum of RAM, you can also try changing both lines, or just the MaxFileCache line, to 4096. This gives 4MB of caching, which should still be okay for audio.

Another form of caching, *read-ahead optimization,* can also interfere with data flow. To turn this off, right-click on My Computer > select Properties > Performance tab > File Settings > Hard Disk tab > drag the read-ahead optimization slider all the way to the left (none).

Cleaning and Maintenance. If you want your CD-R drive, and the data you write onto it, to live long and prosper, then you must be stringent about equipment maintenance and cleanliness. CD-Rs have been rejected by CD duplicators because of a single speck of dust that was present on the blank CD-R before it was recorded. When the laser writes data to the disc, the dust speck causes a shadow that prevents the pits from being formed correctly. This results in an uncorrectable error on playback.

Important points to keep in mind when producing CD-Rs:

- Handle blank CD-Rs by the edges. Never touch the recording surface.

- Never set the blank CD-R down anywhere except in its original jewel case or the CD-R recorder's tray.

- Use Dust-Off® or some other compressed air to remove dust from the CD-R and its carrier (if required) before each recording session.

- For the lowest possible error rate, use only the brand or type of blank CD-R media recommended by your drive's manufacturer.

- Choose a recorder that will produce the CDs appropriate for your requirements. Older CD-R units can have small buffers, record only in track-at-once mode, or leave a mandatory muted space between cuts.

- With most media, recording at 2X gives a lower error rate than either 4X or 1X.

- If your recorder takes longer than expected to recognize the inserted blank CD-R, then the laser lens is collecting dust. Clean it with compressed air, or take it to a repair center to have it cleaned. Do not use cleaning CDs; they can knock the laser out of alignment.

- Record the disc in "Disc-at-Once" mode, rather than multisession or track-at-once. This ensures that the greatest number of drives can read your disc.

- Doing crossfades in real time doubles the amount of data crossing the SCSI bus, so when you use them always do a test run first. If the test fails, circumvent the problem by writing a disc image and burning from that.

- Audio CD-Rs recorded on Philips-based stand-alone recorders such as the Marantz 610 and 620 will not play until they are "finalized" or "fixed up." This process writes the final Table of Contents to the CD-R, turning it into a non-recordable CD.

- Do not allow the CD to have prolonged exposure to sunlight. The error rates will increase.

Mastering

The art of mastering, where songs are adjusted for their optimum level, sequence, and tonal balance, is an art form in itself. It is usually done prior to creating a CD, but nowadays, it's just as likely that you'll create the CD first, live with it for a while, take notes on what needs to be changed, then master it before creating the finished product.

Highly paid and technically skilled mastering engineers often make a significant contribution to a song becoming a hit, almost as much as the songwriter, musician, or recording engineer. The mastering engineer makes sure that a tune can translate to a

PRACTICAL FAQS FOR BURNING YOUR OWN CDS

Q: What do different CD-R recording speeds mean?

A: The "speed" rating of a CD-recorder determines how fast it can record or play back data. Ratings such as "1X," "2X," "4X," and "8X" define multiples of the original speed of first generation CD-ROM players. These data transfer speeds translate to about 150K per second (1X), 300K per second (2X), 600K per second (4X), and 1200K per second (8X), etc.

Q: What are "Disc at Once," "Track at Once" and "Multisession" writing modes?

A: *Disc at Once* is a writing mode that requires the data on a CD to be continuously written, without any interruptions. All of the information is transferred from hard disk to the CD in a single pass, with the lead-in, program, and lead-out areas being written in a single event. There are some CD-R drives that don't support disc-at-once recording.

Track at Once is a writing mode that allows a session to be written as a number of discrete events called "tracks". With the help of special software, the disc can be read before the final session is "fixed" (a process whereby the disc's overall lead-in, program data, and lead-out areas are written), allowing all of the data contained on the disc to be read by any CD or CD-ROM drive.

A *Multisession* disc differs from Track at Once in that several sessions can be recorded separately onto a single disc (with each session containing its own lead-in, program data, and lead-out areas). Thus, data can be recorded onto the free space of a previously recorded CD. (Note that some older CD readers can read only the first session of a multisession disc.) A CD-R that supports multisession can record a number of discrete sessions onto a single disc, and a CD-ROM drive that supports multisession can access the data that was written within any of these sessions.

Q: Can CD-R discs be used as masters for stamping?

A: Yes. If you're planning to do this, first speak with the mastering facility about its specific needs. This facility will help you ensure that no "uncorrectable" or E32 errors are present on your disc. This is important because the software used in the glass mastering process to drive the recorder's laser beam is often set up to terminate the mastering process upon encountering an E32 error. Traditionally, the mastering facilities will require that you record the disc in the "Disc at Once" mode, because it eliminates the linking and run-in and run-out blocks associated with multisession modes. These often are interpreted as uncorrectable errors during mastering.

Q: What is "digital audio extraction"?

A: Certain hardware/software packages can transfer CD audio tracks directly to your hard disk as a 44.1kHz .WAV or AIFF file (depending upon your computer operating system).

Q: What are the options for CD-R printing?

A: CD-Rs can be labeled in several ways, depending on what level of professional look you want and what you can afford.

(continued on next page)

- Using a felt-tip pen is the easiest and fastest way to label a CD-R. However, you should never use a permanent marker pen that contains a solvent, since it can actually permeate the disc surface and cause damage to either the reflective layer or dye layer below the surface. Two recommended pens are the Sanford Sharpie Ultra Fine and the Dixon RediSharp Plus.

- The next least expensive alternative is stick-on labels. However, if you misalign the label or don't smooth the label down and there are air bubbles under the surface, your CD-R runs the risk of spinning out of balance, which could cause reading and tracking problems. Several companies have created label-positioning devices that claim to solve the above problems. Stick-on labels are one of the least expensive professional-looking labeling methods, but be aware that some adhesives can "outgas" over time, and, over the long haul, these solvents can adversely impact the disc. Sticky labels are OK for references, but are not recommended for masters or archival copies.

- Special ink-jet labelers can print four colors onto a custom-faced disc directly from your PC. This is a great option for those who burn lots of discs that have to have a professional look and feel.

- Finally, companies exist that can custom silk-screen your discs. This method uses the same process as mass-replicated discs and only makes sense when dealing with quantities of 100 or more. Note that it takes time to have them printed, and this is the most expensive "personalized" option. But the results are crisp, clean, and professional.

Q: Are there web sites where I can find out the latest info?

A: OSTA (www.osta.org) is an international trade association that's dedicated to promoting use of writable optical technology for storing computer data and images. The www.cdarchive.com/info site provides the latest data and links you to other related sites. Also, the www.adaptec.com/support/cdrec/bufunder.html site includes detailed facts on buffer underrun problems and solutions.

variety of playback systems, and can often correct for problems that creep into the recording process (such as mixing in a room with bad acoustics that leads to mixing with, perhaps, too much bass).

Several programs are designed to help with the mastering process: These are often "plug-ins" that work in conjunction with a "host" digital audio editing program. The host program can't be all things to all people, nor can it accommodate all needs. Plug-ins can fill the gap—they're like accessories for a car, such as a pickup hitch (which is crucial for some people and useless for others). Some mastering tools are available as rackmount boxes that would, for example, insert between a mixer output and DAT, or include a digital interface that allows for doing a DAT-to-DAT digital transfer while adding particular types of processing (such as equalization or compression).

One important element of mastering is limiting a tune's dynamic range so that it is perceived as being louder (when a commercial comes on TV that seems far louder

CASE HISTORY: BURNING A CD
WITH TOAST CD-DA*

The file you want to burn needs to be in one of three digital audio file formats: AIFF (Macintosh native format), .WAV (Windows native format), or Sound Designer II (SDII) format. The latter is the file format used by Digidesign's popular Macintosh sound editing program.

When the file is ready, it's time to start burning. With Toast CD-DA's simple user interface (see Figure 2), just drag and drop your files into a list, enter a little ancillary information called *subcode data*, and cook yourself up a CD.

		Pause	Start		Title	Length	Gain	Xfade	CP PE	ISRC
▷	1	00:02:00	00:02:00		HelloWorld	03:33:08	0.0	⟩⟨|	■ □	
▷	2	00:00:00	03:35:08		Huh?	00:33:43	0.0	I	■ □	
▷	3	00:00:00	04:08:51		TheMule	03:37:00	0.0	]⟨	■ □	
▽	4	00:01:00	07:46:51		Entropy&Inertia	03:23:19	0.0	]⟨	■ □	
			07:45:51		Pause Start					
			07:46:51	1	Track Start					
			11:09:70		Track End					
▷	5	00:02:00	11:11:70		YouDriveLikeAMoron	02:37:56	0.0	I	■ □	
▷	6	00:00:00	13:49:51		Reboot!	02:02:06	0.0	I	■ □	
▷	7	00:00:00	15:51:57		40MinutesOf...	00:35:41	0.0	I	■ □	
▷	8	00:00:00	16:27:23		Attitude	00:30:00	0.0	I	■ □	
▷	9	00:00:00	16:57:23		AlienChaseScene	04:28:00	0.0	]⟨	■ □	
▷	10	00:02:00	21:27:23		Groove31	00:42:63	0.0	⋁	■ □	
▷	11	00:00:00	22:10:11		Dirige	04:09:57	0.0	]⟨	■ □	

Window title: **Erratica** — Add Track / Remove Track / Check Speed / Write Disc — `02 : 0 : 00:00:00`

75 frames/sec ▼ | Absolute Index Times ▼ | Offsets: 0/0 | 13 Tracks 31:05:74

② *Toast CD-DA's streamlined interface makes burning a CD fairly straightforward. When you drag audio files of the songs into the window, the program automatically displays their lengths and start times in minutes, seconds, and frames. To add space between songs, you adjust the number in the Pause column. "Xfade" lets you overlap adjacent songs and perform a volume crossfade between them (you can specify the length of the fade in a pop-up window). The CP and PE checkboxes control whether copy-protect and pre-emphasis codes are placed in the track; the ISRC identification code was not used in this instance.*

Subcode Information. A few bits in each digital word on an audio CD are dedicated to carrying subcode information (see Figure 3). This includes:

• The current running time within each track

• Running time for the entire disc

• Track and index numbers

• Basic table of contents

• Level of copy protection

---

*Now Adaptec Jam.

(continued on next page)

- Whether pre-emphasis (a close-to-obsolete noise reduction technique) is being used
- The Universal Product Code (UPC, which is important if you want to sell your CD through mass distribution channels)
- International Standard Recording Code (ISRC, which states the track's country of origin, owner, year of release, and serial number). Apparently, the ISRC isn't used all that much, so you needn't pay it much attention.

Now that you've literally etched your music into the world's most popular hi-fidelity playback format, evaluate your work. After months of working on each song individually, hearing a project as a whole is a new experience. Some song-sequencing problems are subtle—sometimes the juxtaposition of the chords in two consecutive songs can sound slightly off—and others hit you right in the face. Make sure that the songs flow well from one to the next, without jarring differences in sound quality.

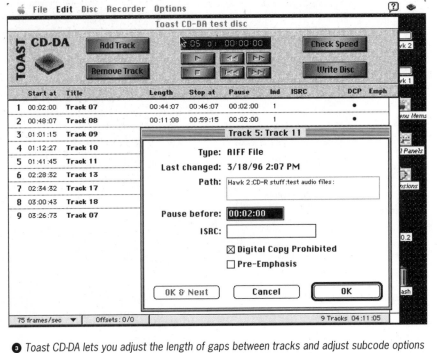

● *Toast CD-DA lets you adjust the length of gaps between tracks and adjust subcode options like pre-emphasis and copy protection.*

than a movie's dialog, you're hearing limiting at work). It may not make sense to squash your music's dynamic range, but just because the CD can reproduce a wide dynamic range doesn't mean that the listening environment can handle it. Many people listen to music in their cars, and a song's quiet parts can get lost in the road noise. Similarly, on the dance floor or on the radio, a song that is not as loud as others is perceived as "weaker."

One common mastering tool, the Waves L1 Ultramaximizer plug-in, isn't used solely for music but for multimedia and game applications as well. This "look-ahead" limiter analyzes a signal and reduces the dynamic range so the whole track can be turned up louder (see Figure 4). The loud parts are still loud, but the quiet parts are louder than they were. The overall mix doesn't sound any different, just more energetic; in short, it kicks (although if you over-limit, it also annoys, and leads to listener fatigue).

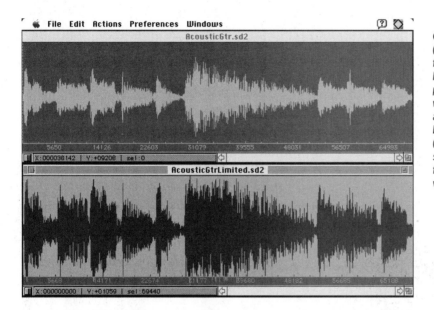

❹ *In the original guitar recording (top window), the average level is fairly low, and there are wide level variations. After the file was processed with Waves' L1 software, the spikes were tamed, allowing a significant overall level boost without adding distortion (bottom window). These screen shots were made in BIAS Peak for the Mac, though L1 runs on a variety of platforms.*

Some programs include a multiband dynamics processor that divides a signal into separate frequency bands, and compresses each band independently. This kind of tweak requires a lot of practice and experience to use effectively. In general, it's best to use DSP tools sparingly, and let your ears be your guide—assuming, of course, that you have well-educated ears. Otherwise, it's probably best to entrust your music to a professional mastering engineer who can really bring out the best in your work.

Mass Production. Making custom CDs works well for small quantities, but it's time-consuming and relatively expensive. Also, since recordable CD media aren't entirely reliable, you really should test each disc before you try to sell it. Given the prices offered by CD manufacturing houses, it makes a lot more sense to mass-produce if you're making more than 100 CDs. Still, owning a CD recorder with the proper software can save you a few hundred dollars on the cost of limited-run duplication if you have more time than money.

For a successful manufacturing run, it's a good idea to check out the information on the Disc Makers web page at www.discmakers.com. The site includes hints on master tape preparation, answers to frequently asked questions, and templates for artwork to download. They even provide helpful tips for building contacts for distribution and sales. Also be sure to visit www. cd-info.com, a wonderful source of CD-R information and links.

Disc Makers stresses listening to your master submission to make sure you like everything about it. Most problems, they say, result when their customers trust the engineer in the recording studio and don't bother to master the music properly before sending the final product out for duplication. The first time they hear it is when they receive a reference copy from the CD manufacturer. You can expect the manufacturer to make sure the recording fills a CD's entire dynamic range, but they will simply duplicate what they are given, without any mastering. Also, if clicks or false starts haven't been trimmed out, don't expect the manufacturer's engineers to clean them up. Unless you provide explicit instructions about changes you want implemented, the engineers have to assume you want the material duplicated exactly as presented.

It's helpful to list the contents of your master tape as thoroughly and precisely as possible, including track start and stop times and anything that needs to be fixed. If there's a click between two songs and you want it removed, note its exact location. If all the songs are separated by silence but you want particular cuts to butt up against one another on the final disc, tell the disc manufacturer not only the song titles, but their exact lengths and even what the audio is doing at that point (e.g., "snare drum' reverb fades to zero"). The better the documentation, the more likely you are to get what you expect. Duplicators often complain that timings from home and project studios are never right; do your best to prove them wrong.

Disc Makers formats their CDs using a Sonic Solutions system. They capture the master tape or home-burned CD digitally, convert the sample rate if necessary to 44.1kHz, make any explicitly-requested edits, then enter subcode data. Then they burn a CD just like the one you pulled out of your CD recorder, which goes to the customer for reference. If the reference isn't right, they do it again (and charge extra). Once the customer approves the reference disc, they make molded plates from it, stamp out the discs, and ship them to your doorstep within a couple of weeks.

It may seem silly that you provide a finished, mastered CD and they recapture it and give it back to you as a reference disc. Assuming you have a cool piece of CD-burning software, you can probably bypass this step. The upside: You save a couple hundred bucks. The downside: You take a crucial step in the production process upon your own shoulders and eliminate one opportunity to catch errors before they cost you a lot more dough. Unless you like to gamble, always ask for a reference disc.

If you have any doubts about your ability to format a disc for commercial use, it may be cost-effective to use the manufacturer's mastering service. You'll pay a few hundred dollars, but gain from the expertise of people who have mastered many musical projects. Then again, no one cares about your music more than you, so if you're ready to climb the relatively steep learning curve required to master music with technical accuracy and artistic finesse, go for it.

The Afterburn

It's easy to forget that your final product is more than music. Without a package that features eye-grabbing artwork, your babies will probably just sit on a shelf. Many musicians have great ideas about packaging, but lack experience turning a visual concept into a finished product. Most CD manufacturers are associated with graphic artists, either in-house or independent, who understand how colors translate from a computer screen into print, what screen resolutions work best with various printing processes, and other arcane matters. If you decide to use these services, again, document everything you want done as clearly as possible and exercise the patience necessary to answer questions thoroughly. The less guessing the graphic artist has to do, the better the odds that the package design will reflect your artistic sensibility.

And don't forget to trademark your band name and copyright your compositions. Trademark searches can be expensive, but there are ways to work around the big legal fees by doing your own search. In the U.S., contact the Federal Patent and Trademark Office at www.uspto.gov/ for information. For copyright forms and information, call the Copyright Office Hotline at 202-707-9100 or check out their web page at http://lcweb.loc.gov/copyright/. To apply for a UPC code that distributors and retailers can use with their barcode readers, call 937-435-3870 or see the UC Council site at www.uc-council.org/.

BY THE BOOK

In 1983, Sony and Philips presented the first specification for CD-Audio to the ISO (International Standards Organization). Being enclosed in a red binder, the spec became known as the Red Book standard. The more technical name for the Red Book spec is ISO 9660.

To differentiate it from audio, the first CD-ROM spec was bound in a yellow cover and became known as Yellow Book. Then came Green Book (CD-i), White Book (Video-CD), Pink Book (Photo-CD), and finally Orange Book (CD-R and MO-R). A few other variations have cropped up to deal with technologies like multisession recording (the ability to stop and restart a writing session), enhanced CD (in which CD-ROM data is added at the end of an audio disc rather than writing it in the first track), and packet writing (enabling data to be written in small increments rather than all at once).

Troubleshooting   7

hen you hook up a bunch of equipment to a computer, there's no guarantee that it will all work perfectly the first time you turn it on. When making music with technology, Murphy's Law ("If anything can go wrong, it will") prevails. The question is, when something doesn't work, what do you do about it?

- Don't panic. Trace your signal flow, both audio and MIDI, from point to point, and think carefully about what's *supposed* to happen. Make sure all of the relevant devices are turned on, their audio outputs are plugged into the mixer, and their volume knobs turned up. Check that the MIDI cables are hooked up properly, with each Out or Thru connected to the next In in the chain. Is everything on the right MIDI channel?

- With software, investigate the status of the relevant drivers (PC) and extensions (Macintosh). Examine every item in the Options and Preferences boxes. To give just one example of a possible snafu, MIDI may be arriving at the computer's hardware input but never reaching the sequencer software because you haven't selected the hardware input as a MIDI source. If all else fails, try de-installing other software (and hardware) that you're not using at the moment: Simplify the system until it works, and then re-install one piece at a time until it breaks. This key technique in troubleshooting will tell you which item (or conflict between items) is causing the problem.

- Read the manual, but even more importantly, check for any README files that may come with the distribution disk or CD. Manuals are printed before a product is shipped, and unexpected bugs and conflicts generally don't appear until after the product is out in the world. The README files often contain invaluable information gleaned from real-world experience; reading these can save you many hours of frustration (or time spent waiting on the line while calling tech support).

- Scan the classified ads in magazines like *Keyboard*. Third-party manuals, videos, and instructional books are available for some of the more popular synthesizers and computer programs.

- When all else fails, call the manufacturer. Unfortunately, tech support hot lines are often swamped, so be ready when you get through. Be sitting at your equipment, with everything turned on. Try everything you can think of before you call, so as not to waste the tech's time. Describe the problem in as much detail as you can—and make sure you understand the answer before you hang up. When they're in a hurry to answer more calls that are stacked up, techs will sometimes try to give you a quick answer that doesn't actually apply to your situation. Don't expect the tech to instruct you in, for example, MIDI basics; they will usually provide information only on their own products.

And do try to be polite with them, no matter how frustrated you get. The tech didn't design the product, and isn't responsible for its failure to perform as you anticipated. If you're dissatisfied, though, either because the owner's manual doesn't explain things clearly or because the product won't do what you need it to, it's perfectly okay to *respectfully* tell the tech what you think. Part of their job is passing on complaints so that the next generation of products can be improved.

- Be patient. There isn't anybody in the business, no matter how expert, who hasn't been mystified on occasion because of strange malfunctions and odd phenomena. There's always more to learn.

Noise

Eliminating noise from home recordings is a many-faceted challenge. Microphones can pick up passing cars or planes. Older components and deteriorating cables can degrade system performance. And then there's the ever-present hum from AC lines.

Noise, which we'll define broadly as unwanted sound, may consist of non-periodic sound waves (e.g., hiss, rumble, clicks) or periodic waves (e.g., hum, interference from monitors, etc.). It can manifest as a distorted or grainy sound, or as artifacts such as clicks, pops, chirps, or warbles.

In whatever form, noise is perhaps the most pervasive, aggravating, and puzzling problem in any recording situation.

Noise can have many causes. For instance, digital devices might drop a few bits here and there, creating strange "ripping" sounds. Then there's acoustic noise (such as sound waves that come from the computer's fan, a creaking floorboard, or passing traffic) which travels through the air to your mic. In contrast, electromagnetically induced noise (electromagnetic interference, or EMI) starts not as sound but as electromagnetic waves, which silently assault audio signals as they're traveling down a wire—with unpleasantly audible results. The ugliest of all noise, however, is digital distortion. This can happen while recording, if an error occurs during file backup, or during signal processing—even when the goal of the processing is to remove other types of noise.

Digital Audio Dropout Culprits: Disk Caching and Thermal Recalibration

Digital sound files are big. When you're recording multiple tracks, or playing several tracks while recording another one, the sheer bulk of data can be too much for your computer's hardware or software. A momentarily overwhelmed computer may drop bits of audio that it should be recording. On playback, the resulting discontinuities usually sound like clicks or pops. This situation is most likely to arise if you're using *disk caching,* a system that saves data in RAM and then writes it to disk in large chunks. During the writing process, the computer may be too busy to store incoming audio data. This is why it's usually recommended that you shut off disk caching during a recording session.

Many older drives also take time out to do a "thermal recalibration," which adjusts the positioning of the drive's read/write head to compensate for the slight expansion of the drive's platter as it heats up. Though thermal recalibration takes only a matter of milliseconds, it can cause noticeable audio dropouts, or even lock up your recording software. Thermal recal dropouts are an annoyance during playback, but during recording, they can destroy a great take. Some manufacturers build drives that will not interrupt read or write operations by postponing thermal recalibration until idle periods. These are given the designation "A/V" drive, which indicates suitability for audio/video applications.

You may need a faster hard drive, one that postpones thermal recalibrations, or a computer with a faster CPU or bus. Before you get out your checkbook, though, learn all you can about optimizing the operating system and digital recording software. Under Windows 3.1, for instance, your software may run better in Standard mode than in 386 Enhanced mode. Or vice-versa. It's also important to make sure you're using up-to-date drivers (which are usually free for downloading from the Web) for your drives and SCSI card. For example, if your hard drives are using MS-DOS drivers under Windows 95, you've taking a major performance hit. The MS-DOS drivers can only move 16 bits of data at a time, while newer 32-bit drivers move twice as much data.

Acoustic Noise

Once you've been working with computers for a while, your mind begins to filter out familiar sounds like whirring fans and clicking hard drives. But your microphones don't, and your music can come out sounding like it was recorded live on a runway at the local airport. No wonder: Some computers have as many as four "propellers" going all the time, such as fans for the CPU, the power supply, any external drives, and maybe even some add-in cards.

Proper placement of directional microphones can help reduce pickup of unwanted sound. And, of course, computer noise won't be a problem if you have an acoustically isolated control booth. That's not the case with most home studios, though.

The simplest, although crudest, solution for home miking (aside from remembering to close the window while you sing or play) is a good noise gate (see Figures 1 and 2). Some multi-effect devices offer a gate algorithm. A gate won't eliminate noise that bleeds through while you're singing, but it may clean up the gaps between

❶ A noise gate, such as this dbx 363X, mutes the signal fed through it when the signal's level falls below a preset threshold. This is a quick way to remove hiss between songs or noise entering a microphone between vocal phrases.

❷ Single-ended noise reduction units, such as the Behringer SNR-202 Denoiser, remove hiss by automatically reducing the treble in a signal during quiet sections.

phrases. If most of the noise comes from one direction, such as a street, using a cardioid (directional) microphone rather than an omni-directional mic may squash the grunge by a few dB. Recording late at night or early in the morning is another way to avoid traffic noise.

When environmental noise does get onto the tape, you often don't hear it in the mix unless you solo the vocal track. The rest of the music will often "mask" the offending noise.

Before recording a track that you may want to keep, try recording just the mic output with no music going through it. Take a look at the residual noise waveform on the screen, or listen to it through headphones, to determine whether you have a background noise problem.

One common solution for eliminating fan noise is to isolate the computer (by putting it in an adjacent room, for instance), place the screen and keyboard in the control room, then use special extension cables to connect the peripherals to the computer. Even placing the computer under a table, and attaching sound-absorbing material under the tabletop, will help a little. However, don't put the computer in a closet or any kind of enclosure without adequate ventilation, as components (particularly the power supply and processor) can overheat and fry.

PC users have another option: a very-low-noise computer like the Quiet PC from Decibel Instruments. The Quiet PC has just one ultra-quiet low-power fan; to compensate for the reduced ventilation, the power supply incorporates extra heat sinking. Also, the Quiet PC's hard drives are encased in a rubber-like material and have special heat sinks. Unfortunately, the Quiet PC costs considerably more than a comparable "noisy" PC, and there's no "quiet Mac" yet.

Another way to eliminate computer fan noise is to turn off the computer, then record your acoustic tracks direct to DAT tape or to a digital multitrack tape deck. Tracks can then be transferred digitally to some computer-based recording systems with no loss of fidelity (see Chapter 4, "Digital to Digital Transfers").

Once you've dealt with the noise-making parts of the computer, you can use insulation (such as fiberglass or acoustic isolating foam) in or on the walls to prevent other noises from leaking into your studio. Other devices that address acoustic noise include acoustic tiles, baffles, and traps, which alter the reflective/absorptive qualities of walls and ceilings. But before you nail up 200 empty cardboard egg cartons on the walls, consider the distinction between soundproofing and sound diffusion. The only way to stop sound from passing *through* a wall is with mass; cement blocks are ideal, and egg cartons will have little effect. What egg cartons do is create a non-reflective surface *inside* the room, which may help give the room a flatter frequency response and a lower overall level of reverberation. Hanging heavy blankets or rugs several inches away from the wall can help reduce sonic reflections as well. For further reading, check out *How to Build a Small Budget Recording Studio from Scratch,* by F. Alton Everest and Mike Shea, or *Building a Recording Studio,* by Jeff Cooper.

Electromagnetically Induced Noise

Electric guitarists know that a guitar will hum when held near a transformer. The coils in the guitar's pickup act as an antenna, picking up electromagnetic waves created by the AC current. The coils convert those waves into a voltage, which the guitar amp turns into hum.

Electromagnetic waves can induce currents in your audio cables, too. Radio frequency (RF) waves from power supplies and monitors can travel through the air for many yards to infect cables. House current produces lower frequency waves (60 cycles per second in the U.S.) that can usually travel only a few inches through the air. But a few inches is more than enough if the house current gets into an audio cable.

Here are some straightforward precautions you can take against electromagnetic noise:

- Don't run audio or digital signal cables next to power cords or transformers ("wall warts"). Audio cables can pick up hum. Even digital audio signals can be affected, though this is less likely. Some experts even avoid bunching audio cables together. If you must place an audio cable next to a power cord, make sure it crosses the power cord at an angle rather than running parallel.

- Look out for "unterminated" audio cables that don't connect to anything on one end. They have a particular tendency to act as antennas. If there's nothing connected to one end of a cable, unplug it completely.

- Keep your audio cables as short as possible: Short cables pick up less RF energy, and what they do pick up is mostly in the higher frequencies. A two-foot length of cable may pick up only noise frequencies that are too high, with levels that are too low, to cause any problems.

- Avoid standard fluorescent lighting in your studio (fluorescents produce considerable acoustic noise, too). Dimmer switches are also problematic. However, compact fluorescent lights are acceptable.

- Try not to connect heavy appliances, such as refrigerators and air conditioners, to the same circuit as audio equipment. The heavy current they draw makes them potent noise sources. Smaller appliances with motors (blenders, hair dryers) are also offenders.

- Try moving a sound card to a different slot in the computer. The inside of a computer is full of electromagnetic noise, and the sound card may just be in a bad spot.

Tracing Hums and Whines. Okay, what if you've taken these precautions and still hear a hum or whine? Usually, one cable, power cord, connector, or piece of equipment is responsible, but it can be hard to find out which one, because EMI tends to spread throughout the system: A computer monitor may produce RF interference, for instance, that is picked up by every cable in the system. Similarly, a single bad connector may cause 60-cycle hum in multiple cables and devices.

You may be able to eliminate RF interference just by reorienting the source—turning your computer monitor 90 degrees, for example—and/or moving it farther away from the "antennas." If reorienting doesn't help, you may need to adjust, repair, or replace the offending piece of equipment

Luckily, RF problems are relatively rare. AC hum, on the other hand, is as common as the proverbial cold.

Symptom of a Blocked Ground Path. Hum can come from poor grounding (see "Hums and Ground Loops," below). If you run short of outlets, you can plug a computer or sequencer into a different outlet than the synths and mixer without creating ground problems, as long as the only connections between the two components are MIDI cables. Theoretically, MIDI cables aren't supposed to make a ground connection—but never assume anything (the cable or MIDI connector may be defective); check those connections if all else fails.

What can clog up a ground path? The list is long: A bad connector, improper house wiring, corrosion on the third pin of a power cord or in the wall socket, or a power cord that has no third pin are likely candidates.

In Figure 3, having equal, low-resistance paths to ground for both the mixing board and the computer should result in a quiet, safe system. On the other hand, if corrosion in the mixing board's plug or socket raises the resistance of the ground path, current flowing through this resistance will create a voltage; given that many components in an audio system connect to ground, this voltage could end up catching a free ride to your signal path. The worse the corrosion, the greater the resistance, and the greater the chance for hum. What's worse, this corrosion may take on the electrical properties of a diode, which can turn radio frequencies into audio frequencies (this is the principle behind the crystal radio) and cause even more interference.

Of course, this assumes all your equipment combined draws no more current than can be supplied by one circuit—a fair assumption for most home studios. If you're concerned about how much current your equipment draws, here's how to figure it out: Look on the back of each device to determine its wattage (this is usually printed on a plate or label next to where the AC cord enters the unit), then use the formula *amps = watts/volts* to figure out its amperage. For instance, a 25-watt device running on a standard 120V circuit will draw 25/120, or about 0.21 ampere. Add up the total amperage of your system, and compare this figure with the ampere rating of your breaker box's circuit breaker.

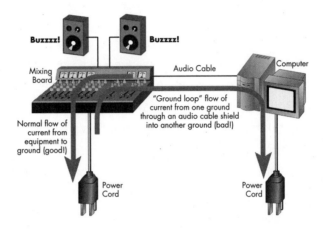

❸ *When an electrical signal has several possible paths to ground, there can be voltage differences between these paths. This can allow for the current to flow in the system's ground connections, resulting in the dreaded 60-cycle hum.*

Reassemble your complete audio system one component at a time. If hum suddenly appears, look at the most recent piece of equipment you installed, its power cord, audio cables, and connectors. If one particular analog audio cable is causing problems, you may be able to improve its performance by using a product such as DeoxIT®, from Caig Laboratories, to clean the connectors.

Gain-Staging

Optimizing mixer settings (or more generally, the audio settings in your studio) for the maximum headroom and minimum noise is called *gain-staging*. For starters, set any audio source to produce the maximum possible volume, short of distortion. If it's a synthesizer, crank the volume knob all the way up. Then use your mixer's input trim controls to reduce the level of the signal in that mixer channel. Turning down the input trim reduces any noise added by the audio cables as well.

If the input trim is set too high, the incoming audio will distort. (Some mixers provide a clipping indicator on each channel so that you can see at a glance where the distortion occurs. If you don't have clipping indicators, check to see whether you can switch your mixer's output metering to monitor one input channel or bus at a time to help isolate the distortion's source.) Conversely, if the input trim is too low, you won't take advantage of the mixer's available headroom, which may add noise to the mix. To set the input trims properly, turn up each one and play the loudest audio that will be handled by that channel; when clipping occurs, turn the trim down just far enough to cause the clipping to disappear, then turn down just a bit more to add a little safety margin.

Noise Prevention and Removal

You cannot completely prevent audio systems from generating and picking up noise. A certain amount of noise, particularly hiss, is inherent in audio equipment.

Detective Work. To analyze the electronic noise in your system, turn up the power amp and master output volume sliders as far as possible. (And be *extremely* careful not to feed in any input signals!) This will magnify the hiss and hum in your system. Next, turn down one mixer channel at a time while listening closely to the noise, or turn them all down and bring up one channel at a time.

If you're able to isolate some especially noisy channel through this procedure, figure out why it's noisy. Is a trip to the shop in order, or is a bad cable perhaps at fault? Check the audio cable's path to make sure it doesn't lie alongside a power cord. Some distortion and chorusing algorithms produce audible noise even when no notes are being played; you may be able to remedy this problem by switching the noisy instrument to a different setting, or bypass mode, when it isn't playing.

Here are more tips to minimize or remove noise:

- Avoid cheap equipment. You can do Windows-based hard disk recording with a consumer-level sound card, but one designed for pro applications will be quieter. Generally, some indications of pro audio gear are balanced instead of unbalanced jacks: AES/EBU instead of S/PDIF for digital signals, and XLR instead of phone jacks for analog signals. (Note that phone jacks are sometimes set up in a balanced configuration, but XLR connectors are almost always balanced.)

- If you're adding MIDI tracks to your recording, don't rely on MIDI volume control messages to mute a channel. MIDI instruments often generate the same amount of hiss regardless of MIDI volume control messages—and even if they don't, the cable between the synthesizer and the mixer could add some noise to the system. Use your console's mute buttons, or automation in your hard disk recording software, to mute the audio channels of unused MIDI instruments.

- You can use noise removal software for the Mac or Windows; some of these programs are available in stand-alone form, while others are plug-ins for "host" programs. To use these, you first take a sample of a track where there is no recorded material—just before the music begins, for instance. Any sound on the track at that point is pure noise. Once that sample has been analyzed, the software goes through the entire track and digitally removes that noise pattern. Use these types of programs judiciously, however, or they may introduce other artifacts such as warbles and chirps.

You can also use hardware or software equalization (EQ) to reduce the level of an offending frequency. For example, if you have a hum or whine at a particular frequency, use the EQ on your mixing board or hard disk recording software to attenuate that frequency. This will obviously affect whatever musical material is in the same frequency range, so the best EQ settings will be those that affect the narrowest possible band of frequencies. Ideally, you'll catch the whine or hum while setting up or doing a test recording, so the noise will never get on your final recording. You may be able to determine the frequency and the amount of attenuation required by a little experimentation.

Digital Clipping

With digital recording, you need to listen to your tracks carefully after recording to make sure that transient peaks didn't cause even momentary clipping.

Clipping occurs when the input signal exceeds the system's maximum headroom. This cuts off the peaks of the input signal, turning rolling hills into mesas (see Figure 4). Analog recorders can also clip, and it sounds bad. With a digital recorder, however, clipping sounds particularly obnoxious.

The most natural way to prevent digital clipping is simply to turn the input signal down until it no longer produces clipping. But in certain situations, such as a live band recording, this may not be practical. If your source material has a wide dynamic range, turning it down also has the effect of reducing the low-level signals to an undesirable level.

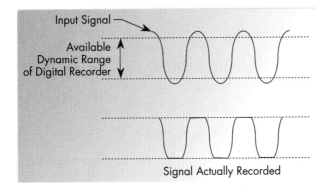

Input Signal

Available Dynamic Range of Digital Recorder

Signal Actually Recorded

❹ Clipping occurs when the input signal (top) has a level that exceeds the maximum dynamic range of the system. The outer edges of the sound wave are "clipped" off (bottom), which drastically changes the harmonic content of the sound. Momentary clipping usually sounds like a pop or click; when a whole area of the wave clips, your music will be inundated by thick, grinding, static-like noise. The solution: Turn down the input until the clipping goes away.

Compression. To prevent clipping while preserving a hotter overall signal level, you can run the signal through a compressor before it hits the recorder. A compressor reduces the signal's level when the level rises above a certain threshold level (see Chapter 4, "Compression"). Typically, you set the threshold level to reduce only the loudest sounds. You also set a compression ratio. For instance, you could reduce all sounds above −10dB by a ratio of 3:1.

However, you need to be aware of clipping on your compressor's inputs. If you're squashing the signal, use the output level control to maintain the desired output level. If you feed too high a signal into the compressor to maintain a loud output, you could overload the compressor's input.

Compression does alter the sound of the material. For instance, rock bands often put heavy compression on drums to "squash" and "tighten up" the sound. If you're going for a natural sound, however, you'll want to keep the threshold relatively high (thus bringing compression into play as seldom as possible) and the compression ratio as low as possible (thus minimizing the effect of the compression when it does come into play).

Limiting. If the compression ratio is infinite, then nothing louder than the threshold will get through the compressor. In that case, the compressor becomes a limiter. Almost all compressors can also function as limiters. Many units add expander, noise gating, EQ, or de-essing functions as well. (De-essing reduces the amount of sibilance in vocal material. Sibilance is a hissing or whistling sound often associated with an "s," "ch" or "sh" sound. Often considered undesirable in vocal material, sibilance can be reduced by cutting down particular frequencies. Some software programs also include de-essing algorithms.)

Quantization Noise

The digital audio equivalent of "jaggies" (in computer graphics, this refers to the choppiness you see in curved lines) is called quantization noise. And just as visual jaggies show up most clearly on fine lines, quantization noise is most apparent with low-level signals. It's also present in louder digital audio, but any noise will be completely masked by the music.

Dithering. Digital audio engineers use a process called "dithering" to minimize quantization noise. Strange as it may seem, dithering involves adding a bit of controlled noise to the audio signal. This smooths out the transitions from one level to another, thus giving a subjectively more pleasing sound. Dithering is often done when converting 20- or 24-bit signals to 16 bit. Just cutting off the last few bits doesn't sound as good as adding dither to the last few bits.

Noise is a beast that can never be entirely tamed, and often comes snarling out of its cage when you least expect it. Defend yourself with knowledge, perseverance, and careful listening. The results—music that shines through clean and crisp—are worth it.

Hums and Ground Loops

What was that buzz? That strange hum? The digital hash from your computer that's showing up in the mic preamp? You may be a victim of *ground loops,* which can occur when using multiple AC-powered devices.

A ground loop means there is more than one ground path available to a device. In the Figure 5, one path goes from device A to ground via the AC power cord's ground terminal, but A also sees a path to ground through the shielded cable and AC ground of device B. Because ground wires have some resistance (the electronic equivalent of

friction), there can be a voltage difference between the two ground
lines, thus causing small amounts
of current to flow through ground.
This signal may get induced into
the hot conductor. The loop can
also act like an antenna for hum
and radio frequencies. Furthermore, many components in a
circuit connect to ground. If that
ground is "dirty," this noise might
get picked up by the circuit.
Ground loops cause the most prob-

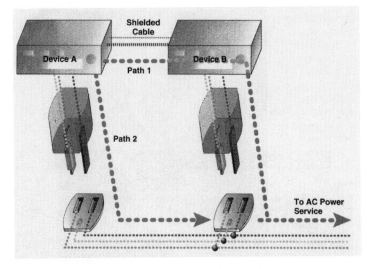

5 *Multiple paths to ground can create ground loops.*

lems with high-gain circuits, since massive amplification of even a couple millivolts of
noise can be objectionable.

There are two main fixes: Break the loop by interrupting the audio ground, or by
interrupting the AC ground line. The preferred method depends on the nature of the
problem, so let's look at various options.

Ground Lifters. Some musicians simply "lift" the AC ground by plugging a 3-wire
cord into a 3-to-2 adapter. This is definitely *not recommended* since it eliminates the
safety protection afforded by a grounded chassis. However, rather than spending
another page or two explaining why you shouldn't do this, just don't do it, okay?

Solution No. 1: The Single Plug. You can solve many ground loop problems by
plugging all equipment into the same grounded AC source, such as a barrier strip that
feeds an AC outlet through a short cord, as this attaches all ground leads to a single
ground point. However, it is crucial that the AC source is not overloaded and is properly rated to handle the gear plugged into it.

Solution No. 2: The Broken Shield. A solution for some stubborn ground loop
problems is to isolate the piece of gear causing the problem, and disconnect the
ground lead (shield) *at one end only* of one or more of the audio patch cords
between it and other devices. The inner conductor is still protected from hum by a
shield connected to ground, yet there's no completed ground path between the two
devices except for AC ground.

Sometimes a ground loop shows up as objectionable only if the grounded metal
chassis of a piece of rackmount gear contacts the metal rail of a rack cabinet. There's
an easy fix: HumFrees, from Dana B. Goods, are little plastic strips that attach to your
device's rack ears and insulate the device from the rack (see Figure 6). They can be
particularly effective with rackmount computer peripherals that dump a lot of garbage
to ground.

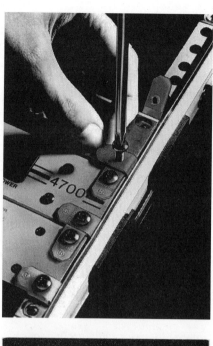

❻ *Humfrees non-conductive bushings (distributed by Dana B. Goods) help break ground loops by isolating rackmount devices from each other.*

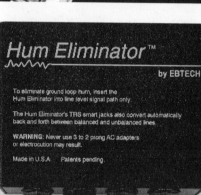

❼ *The Ebtech Hum Eliminator uses audio isolation transformers to break ground loops. The stereo model is shown.*

Solution No. 3: Audio Isolation Transformer.

Using a 1:1 audio isolation transformer is much more elegant than simply breaking the shield, but delivers the same benefit: It interrupts the ground connection while carrying the signal. Although a cord with a broken shield is less expensive, the transformer offers some advantages. If necessary, it can also change impedance or levels if you choose a transformer with different impedances for the primary and secondary windings (e.g., use the transformer to boost the level of a device with a fairly low output; this gives less noise than turning up the mixer's pre-amp gain).

For a commercial implementation, check out Ebtech's rackmount Hum Eliminator. This consists of audio transformers in a rackmount case, and uses TRS (tip/ring/sleeve) phone jacks that work with balanced or unbalanced lines. To "break" an audio ground line, just use one of the transformers in the Ebtech instead. (Ebtech also makes a model that converts back and forth between +4 and −10 signal levels—see Figure 7.)

Solution No. 4: AC Isolation Transformer.

Many times, you can also break a loop by removing the direct connection from a piece of gear to AC ground through an isolation transformer (see Figure 8). This doesn't always work because the ground loop may not involve the AC line but various ground-to-ground connections; however, loops involving the AC line generally seem to be more problematic and common. Breaking the audio connection is a simpler, lower power solution (and can also minimize computer-generated "hash"), but an AC isolation transformer provides ancillary benefits (see Chapter 3, "Studio AC Protection"). In summary, an AC isolation transformer can clean up the AC line, reduce spikes and transients, and provide performance almost equal to that of a separate AC line.

One such device is made specifically for musicians: MIDI Motor's Hum Buster, which has a large transformer with 10 isolated AC outlets.

So which is better, breaking the audio connection or the AC connection? It depends. If you have a lot of microprocessor-controlled gear and less than ideal AC,

❽ The Furman IT-1220 isolation transformer is a new type of power conditioner that provides balanced power, greatly reducing ground-loop hums and radiation from electrical cables into audio circuits.

adding isolation transformers can solve various AC-related problems and get rid of ground loops. If you just have a simple ground loop problem, then patching in an audio isolation transformer may be all you need.

Phase Problems

It is important to make sure your studio connections are *in phase*. But before we discuss detecting and solving phase problems, let's explain the concept of "phase."

Without getting into the mathematical details, in most cases we're really talking about a change in signal *polarity*. Reversing the polarity inverts the entire signal so that the negative-going parts of the waveform become positive-going and vice-versa (Figure 9). In practical terms, at any part of the waveform where the speaker cone would have been moving toward you, flipping the polarity causes the cone to move away from you.

Flipping polarity is independent of the wave's frequency, but there's another form of phase reversal (as used in phase-shifter effects) that *is* frequency-dependent. This creates the phase reversal by delaying the input signal and adding the delayed signal to the original. Most musicians and engineers understand what the term "phase reversal" means, so we'll call it that, even though "polarity reversal" is technically a more accurate term in most situations.

Flipping a signal's phase may or may not mean too much by itself; that's a matter of debate. Some people feel you can definitely hear a difference with instruments like drums. For example, with a real kick drum, the first rush of air pushes out at you. If this signal goes through a system that doesn't change phase, the speaker will push air out to re-create the sound of the kick. But if the signal flips phase, then the speaker will pull in to move the required amount of air. It will still sound like a kick drum, although many people hear a subtle—but important—difference.

❾ An in-phase and out-of-phase signal.

In any event, there's no debate that mixing an out-of-phase signal with an in-phase version of the same signal can cause a weakening and "thinness." Phase problems occur a lot when you're using two microphones, since, depending on their spacing, they can pick up a signal's waveform at different points, thereby creating phase differences at various frequencies. Problems can also occur with parallel effects. For example, if an echo signal is out of phase with respect to an in-phase dry signal and the two are mixed together, the echoed signal will tend to cancel the dry signal somewhat, resulting in a thinner sound.

There are many opportunities for phase problems in the studio. Balanced cables can be miswired, some balanced gear assigns pin 3 instead of pin 2 to the "hot" connection (even though an international standard defines pin 2 as hot), some "vintage" effects weren't too careful about phase, and even some new gear will have a design problem crop up from time to time that flips the phase.

Testing 1-2-3. Phase meters, which can detect an out-of-phase condition, are expensive. Fortunately, two-track digital audio editors make a pretty good substitute. With such software you can determine not only whether a device's output is in phase with its input, but in some cases whether a signal is correctly phased or reversed.

To hook up your test setup, split the input signal and send it both to the input of the device being tested and to the digital audio editor's left channel input. This is your reference. Then feed the output signal of the device being tested to the digital audio editor's right channel. You can split this off to an audio monitor as well if you want to hear what's going on. Record a few seconds of stereo audio, then inspect the waveforms in the left and right channels.

As one example of how to use this technique, here's how you would test a mixer to make sure all outputs were in phase. Patch something like a drum sound generator into the mixer input, then test the output at a variety of points: master out, submaster out, monitor out, sends out, direct out, etc. Figure 10 shows a comparison of the input and the send output; the two waves are in phase. If they were out of phase, the peaks and valleys would have the same shape, but go in reverse directions—in other words, when the waveform rose on the upper channel, it would fall by an equal amount in the lower channel. Figure 11 shows what this would look like.

Vintage guitar effects are notorious for phase problems, and are well worth testing. It's also a good idea to test the entire input-to-speaker chain to make sure nothing's amiss. In particular,

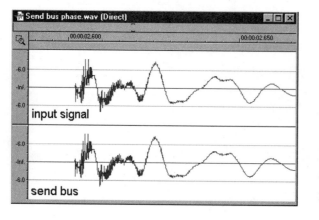

10 *Comparing the phase relationship between a mixer input and the mixer's bus send output.*

check that there isn't a phase difference between the left and right channels, as that can have disastrous results in a mix.

Absolute Phase. Figure 12 shows an example of absolute and flipped phase with guitar. The top channel shows the guitar's original, in-phase signal; note how it starts by going positive. The bottom channel shows the same guitar signal, but flipped in phase. Note how it starts by going negative. It seems that you can identify the absolute phase of many drums similarly—look for an upward slope at the beginning of the signal. However, these are just a few examples; some signals do start off with negative transients.

Fixing Phase Problems. If you feel that phase-flipping does matter, the same program that identified the problem can also provide the solution. Just about all digital audio editors let you select an audio region and reverse the polarity, so you can always import files from your multitrack hard disk recorder (or bounce signals over from digital tape), correct the polarity, then send it back from whence it came. Simple!

In any event, boot up your two-track editor one of these days and take an hour or so to check out the phase integrity of your system. You never know what evil lurks in the wiring of cables.

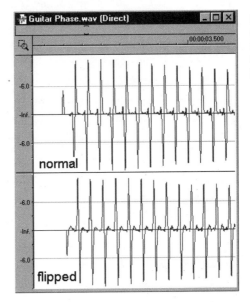

❶ What the bus send output waveform would look like had it been reversed.

❷ Two versions of a guitar signal: in phase (upper) and out of phase (lower).

For Further Reading

Anderton, Craig. *Home Recording for Musicians.* (revised and updated) (Amsco Publishing, 1996.)

Anderton, Craig. *MIDI for Musicians.* (Amsco Publications, 1986.)

Anderton, Craig. *Multieffects for Musicians.* (Amsco Publications, 1995.)

Benford, Tom. *Welcome to PC Sound, Music, and MIDI.* (MIS Press, 1993.)

Eargle, John. *Handbook of Recording Engineering.* (Van Nostrand Reinhold, 1996.)

Holsinger, Erik. *How Music and Computers Work.* (Ziff-Davis Press, 1994.)

Huber, David Miles. *Hard Disk Recording for Musicians.* (Amsco Publications, 1995.)

Luther, Arch C. *Principles of Digital Audio and Video.* (Artech House, 1997.)

McIan, Peter. *Using Your Portable Studio.* (Amsco Publications, 1996.)

Pahwa, Ash, Ph.D. *The CD-Recordable Bible.* (Eight Bit Books, 1994.)

Parker, Dana, and Robert Starrett. *CD-ROM Professional's CD-Recordable Handbook.* (Pemberton Press, 1996.)

Pohlman, Ken. *Principles of Digital Audio.* (second edition, SAMS Publishing, 1989.)

Rona, Jeffrey C. *Synchronization from Reel to Reel: A Complete Guide for the Synchronization of Audio, Film & Video.* (H. Leonard Publishing Corporation, 1990.)

Rubin, David M. *The Desktop Musician.* (Osborne McGraw-Hill, 1995.)

Rumsey, Francis. *The Audio Workstation Handbook.* (Focal Press, 1996.)

Watkinson, John. *An Introduction to Digital Audio.* (Focal Press, 1994.)

Watkinson, John. *The Art of Digital Audio.* (Focal Press, 1994.)

Glossary

ADSR: Attack, Decay, Sustain, Release. An envelope generator configuration.

A/D converter: Analog-to-digital converter, also called an ADC. A device that changes the continuous fluctuations in voltage from an analog device (such as a microphone) into digital information that can be stored or processed in a sampler, digital signal processor, or digital recording device.

AES/EBU: This professional digital audio format is named after the acronyms of the Audio Engineering Society and European Broadcasting Union, two groups that define how the data will be transmitted electrically. Although it travels on a single cable, AES/EBU is a stereo signal and is very similar to the S/PDIF standard (see below). At the physical level, AES/EBU audio uses three-pin XLR connectors.

aftertouch: See pressure sensitivity.

AIFF: Audio Interchange File Format, a standard audio file format supported by Macintosh and some Windows applications.

aliasing: The unwanted frequencies which are produced when a sound is sampled at a rate which is less than twice the frequency of the highest frequency component in the sound. These unwanted frequencies are typically high frequency tweets and whistles.

amplitude: The amount of a signal. It can relate to volume in an audio signal or the amount of voltage in an electrical signal.

analog: A signal or device that varies in a continuous, non-stepped manner.

attack: The first parameter of an envelope generator; determines the rate or time it will take for the event to reach the highest level before starting to decay.

balanced: An audio signal configuration in which two "hot" leads carry audio signals that are identical but of opposite polarity. Balanced connections are good for reducing interference induced into the cable.

bit: A bit is a single piece of information assigned a value of 0 or 1, as used in a digital computer. Computers use digital words which are combinations of bits. A 16-bit word can represent 65,536 different numbers.

bounce: When recording or sequencing, to bounce tracks means to combine (mix) several tracks together and record them on another track.

buffer: An area of computer memory that is used to temporarily store data.

byte: A computer word made up of eight bits of data. See bit.

cardioid: A directional microphone with a heart-shaped, narrow pattern, which picks up from directly in front of the mic and tends to reject signals from other directions. Good for recording drums.

CD-ROM: Compact Disc Read-Only Memory. A laser-encoded disc that can store large amounts of computer data.

cent: Unit of pitch equal to 1/100 of a semitone.

channel, output: The circuitry through which an instrument outputs individual notes.

channel, MIDI: An information pathway through which MIDI information is sent. MIDI provides for 16 available channels, each of which typically addresses one MIDI instrument.

chorus: A voice doubling effect created by layering two identical sounds with a slight delay (20-50 ms) and slightly modulating the pitch of one or both of the sounds.

clipping: Distortion caused by recording at too high a level.

clock: A steady pulse from a generator which is used for synchronizing sequencers, drum machines, etc. Common sequencer timing clock rates are 24, 48, or 96 pulses-per-quarter note. MIDI timing clocks run at a rate of 24 ppq.

Data compression: Data reduction. Used to decrease file size and transfer time.

NOTE: This should be changed to data compression and listed under D, as compression is also an analog audio technique.

crossfade: To gradually fade out one sound while fading in another so that a seamless transition is made between the two sounds.

cutoff frequency: The frequency above which a low pass filter will start attenuating signals present at its input.

cycle: One complete oscillation or vibration of a signal.

DAT: Digital Audio Tape recorder (also refers to the actual tape cassette used by this format, resembling a miniature video cassette). DAT decks can record standard analog signals, but most also have digital inputs and outputs for direct recording of digital audio.

decay: The second stage in an ADSR type envelope generator.

decibel (dB): A reference for the measurement of sound energy. In theory, the minimum change in volume that the human ear can perceive.

delay: A controllable time parameter giving the ability to start an event only after a predetermined amount of time.

depth: The amount of modulation. Sometimes called amount, width, intensity or modulation index.

digital: Voltage quantities or other data represented as binary numbers. Digital information is not audible and so must be converted to analog form by a DAC before it is output.

D/A converter (DAC): Digital to analog converter. A device which interprets digital information and converts it to analog form. All digital musical instruments must have a D/A converter so that we can hear their output.

DMA: Direct Memory Access. With DMA, a peripheral can access memory directly, bypassing the CPU, which saves time and frees up the CPU for other tasks. Different devices are assigned different DMA numbers for identification.

DSP: Abbreviation for digital signal processing.

dynamic range: The range of the softest to the loudest sound that can be produced by an instrument, or playback/recording medium. This can also be the difference between the low and high signal levels obtainable by a velocity sensitive keyboard. The greater the dynamic range, the more sensitive the keyboard.

envelope: A shape representing the changes in a sound's amplitude (or other parameter) over time.

envelope generator: A circuit, usually triggered by pressing a key on a keyboard, that generates a changing voltage (or data string) with respect to time. This voltage typically controls a VCF or VCA. An ADSR is one type of envelope generator. See ADSR.

EQ: Abbreviation for equalization.

equalizer: A device which allows attenuation or emphasis of selected frequencies in the audio spectrum. Equalizers usually contain many bands to allow the user a fine degree of frequency control over the sound.

Fast Fourier Transform (FFT): A computer algorithm which derives the fourier spectrum from a sound file.

filter: A device used to remove unwanted frequencies from an audio signal thus altering its harmonic structure. Low pass filters are the most common type of filter found on music synthesizers. They only allow frequencies below the cutoff frequency to pass. High pass filters only allow the high frequencies to pass, and band pass filters only allow frequencies in a selected band to pass through. A notch filter rejects frequencies that fall within its notch.

flange: An effect created by layering two identical sounds with a slight delay (1-20 ms) and slightly modulating the delay of one or both of the sounds. The term supposedly comes from the early days of tape recording when delay effects were created by pressing on the flanges of the tape reels to change the tape speed.

frequency: The number of cycles of a waveform that occur in a second.

FSK: Frequency Shift Keying. An audio tone (frequency) modulated by a pulse wave, which is used both for data transfer and also for sequencer and drum machine synchronization.

fundamental: The first, lowest note of a harmonic series. The fundamental frequency determines a sound's overall pitch.

gain: The factor by which a device increases the amplitude of a signal. Negative gain attenuates a signal.

green book: The Compact Disc Interactive (CD-I) standard was released by Phillips in 1978 and allows for full-motion video on a standard 5-inch disc. This requires a dedicated CD-I player and is not compatible with an audio CD player.

ground loop: Hum or other noise caused by current circulating through the ground connections of a piece of equipment or system. This is due to different voltage potentials existing along different ground lines.

hard disk: A storage medium for digital data which can hold more information and access it faster than a floppy disk.

hertz/Hz: A unit of frequency equal to 1 cycle per second.

high pass filter: See filter.

IRQ: Interrupt Request Line. In IBM-PCs, when a peripheral device like a soundcard or a CD-ROM drive needs to communicate with the computer's CPU, it sends a signal called an interrupt via a specific IRQ, causing the CPU to stop what it's doing and pay attention. Problems will result if two or more peripherals are set to the same IRQ value.

jack: A (female) receptacle into which a plug is inserted. See plug.

layering: The ability to place or stack two or more sounds on the same keyboard range to create a denser sound.

LFO: Low Frequency Oscillator. An oscillator used for modulation whose range is below the audible range (20 Hz). Example: Varying pitch cyclically produces vibrato.

loading: To transfer from one data storage medium to another. This is generally from disk to RAM memory or vice-versa, as opposed to saving from RAM to disk.

looping: Looping is the process of repeating a portion of a sample over and over in order to create a sustaining or repetitive sound. The looped sound will continue as long as the key is depressed. A sound is usually looped during a point in its evolution where the harmonics and amplitude are relatively static in order to avoid pops and glitches in the sound.

lossless compression: A technique that reduces the size of a file without sacrificing any of the original data. (When expanded, the compressed file becomes an exact replica of the original file.)

lossy compression: A technique in which some data are deliberately discarded to reduce the size of the file. (When expanded, the compressed file is lower in quality than the original file.)

low pass filter: A filter whose frequency response remains flat up to a certain frequency, then rolls off (attenuates signals appearing at its input) above this point.

memory: The part of a computer responsible for storing data.

merge: Combining sequences, sounds, tracks, MIDI data, etc.

MIDI: Acronym for Musical Instrument Digital Interface. MIDI enables synthesizers, sequencers, computers, rhythm machines, etc. to be interconnected through a standard interface. MIDI is an asynchronous, serial interface, which is transmitted at the rate of 31.25 KBaud or 31,250 bits per second.

MIDI clock: Allows instruments interconnected via MIDI to be synchronized. The MIDI clock runs at a rate of 24 ppq.

MIDI continuous controller: Allows continuously changing information such as pitch wheel or breath controller information to be transmitted over a MIDI cable. Continuous controllers use fairly large amounts of memory when recorded into a MIDI sequencer. Some standard MIDI continuous controller numbers are listed below:

PWH = Pitch Wheel

CHP = Pressure

1 = Modulation Wheel

2 = Breath Controller

3 = (Pressure on Rev. 1 DX7)

4 = Foot Pedal

5 = Portamento Time

6 = Data Entry

7 = Volume

8 = Balance

10 = Pan

11 = Expression Controller

16-19 = General purpose controllers 1-4 (High Res.)

64 = Sustain Switch (on/off)

65 = Portamento Switch (on/off)

66 = Sustenuto (chord hold)

67 = Soft Pedal (on/off)

69 = Hold Pedal 2 (on/off)

80-83 = General purpose controllers 5-8 (Low Res.)

91 = External Effects Depth

92 = Tremolo Depth

93 = Chorus Depth

94 = Detune

95 = Phaser Depth

96 = Data Increment

97 = Data Decrement

mini-phone: An ⅛" diameter connector found on many portable tape players and computer audio setups, identical to the phone connector but smaller and shorter. See phone.

modulation: The process of one audio or control voltage source influencing a sound processor or other control voltage source. Example: Slowly modulating pitch cyclically produces vibrato. Modulating a filter cyclically produces wa-wa effects.

monophonic: A musical instrument that is capable of playing only one note at a time. Music with only one voice part.

multitimbral: The ability of a musical instrument to produce two or more different sounds or timbres at the same time.

multitrack: Recording a musical piece by dividing it into tracks, and combining the tracks during playback.

normalize: A digital processing function that increases the amplitude of a sound file until the peak amplitude of its loudest sample reaches 100% of full scale.

orange book: The orange book defines the standard for writable or recordable media such as CD-Rs and magneto optical discs. It defines where the data can be written and, in the case of the MO, how it is erased and rewritten.

OMS: Open MIDI System. OMS acts is a central MIDI driver between OMS-compatible hardware and software. Created by Opcode Systems.

panning: To move an audio signal from one output to the other. Panning a sound between two speakers changes the apparent position of the sound in the stereo field.

patch: Referring to a particular sound created on a synthesizer. Comes from the use of patch cords on the original modular synthesizers.

phone: A ¼" diameter connector, a.k.a. "guitar cord." Phone plugs and jacks come in both mono and stereo versions.

plug: A (male) connector that has one or more protruding pins and fits into a jack. See jack.

Plug-In: Software that extends the functional capabilities of a host program.

polyphonic: A musical instrument that is able to play more than one note at the same time. Music with more than one voice part.

preset: A preprogrammed sound and control setup on a sampler or synthesizer. Presets can be made up in advance of a performance, stored in memory, then recalled instantly when desired.

pressure sensitivity: The ability of an instrument to respond to pressure applied to the keyboard after the initial depression of a key. Sometimes called aftertouch.

Program Change: A MIDI message that tells a synthesizer to change from one instrument sound to another.

proximity effect: When cardioid microphones

are placed very close to the sound source, a boosting of the bass frequencies occurs which is known as the proximity effect. Mostly associated with dynamic microphones.

punch-in: When recording, punching in overwrites a previously recorded track starting at the punch in point.

punch-out: When recording, punching out stops the recording process started by a punch in, thus preserving the previously recorded track starting at the punch out point.

Q: The figure expressing a filter's resonance. Varying Q varies the sharpness of the filter sound.

quantize: Correcting rhythmic irregularities in sequenced music by moving notes to, or closer to, the nearest division of a beat.

RAM: Random Access Memory. The memory in a computer in a computer that stores data temporarily while you are working on it. Data stored in RAM is lost forever when power is interrupted to the machine if it has not been saved to another medium, such as floppy or hard disk.

RCA: A round single-pin connector with a protruding sleeve, commonly used in consumer audio gear. Occasionally referred to in consumer electronics circles as a "phono" connector.

red book: 16-bit, 44.1kHz audio. Red book is the prerecorded CD audio standard that you find in music stores today. Because of this standard, any CD will play in any audio compact disc player. Specified are the sample rate (44.1 kHz), type of error detection and correction, and how the data is stored on the disc.

release: The final part of a sound's envelope, when the amplitude returns to zero.

resonance: A frequency at which a material object will vibrate. In a filter with resonance, a signal will be accentuated at the cutoff frequency. See Q.

reverb: An audio effect that recreates multiple sound reflections in various acoustic environments.

sample: A digitally recorded sound.

sample rate: When digitally sampling a signal, the rate at which level measurements of the signal are taken. The higher the sample rate, the higher the sound quality.

sampling: The process of recording a sound into digital memory.

SCSI: Acronym for Small Computer Systems Interface. An industry standard interface that provides high-speed access to peripheral devices such as hard disk drives, optical discs, CD-R drives, etc.

sequencer: A device which steps through a series of events. A digital sequencer may record keyboard data, program changes, or realtime modulation data to be played back later, much like a tape recorder or player piano. Digital sequencers use memory on the basis of events (key on, key off, etc.) while a tape recorder uses memory (tape) on the basis of time.

serial interface: A computer interface in which data is transmitted over a single line one bit at a time. The MIDI interface is an example of a serial interface.

signal processing: Using electronic circuitry to modify a sound.

signal-to-noise ratio (SNR): The SNR, measured in decibels (dB), is a way of describing how loud the signal (i.e., the music) is, compared to the residual noise (static, hiss) in the recording.

SIPP: A RAM expansion module similar to a SIMM, but less common.

SMPTE: Acronym for Society of Motion Picture and Television Engineers who adopted a standard time code in order to synchronize video and audio. SMPTE information is in the form of hours, minutes, seconds, and frames. There are two types of SMPTE time code, Longitudinal Time Code (LTC), which is typically recorded on audio tape, and Vertical Interval Time Code (VITC), which is often recorded on video tape.

software: The programs or sets of instructions describing the tasks to be performed by a computer.

Song Pointer: MIDI information which allows equipment to remain in sync even if the master device has changed location by causing auxiliary gear to autolocate to this message. MIDI Song Pointer (sometimes called MIDI Song Position Pointer) is an internal register (in the sequencer or autolocator) which holds the number of MIDI beats since the start of the song.

sound module: A synthesizer or sampler without an attached keyboard. Produces sounds via input from an external MIDI device.

S/PDIF: A consumer version of AES/EBU stereo digital audio. The acronym stands for Sony/Philips Digital Interface Format. S/PDIF (pronounced "spih-diff") is carried on unbalanced

RCA connectors, which are sometimes referred to in this context (incorrectly) as "coaxial" connectors. S/PDIF signals can also be transmitted on optical cables. **NOTE: Coaxial simply means shielded cable, I believe...**

split: Dividing the range of a MIDI keyboard or other controller into different sections, each of which controls a different instrument or sound. Also the point at which this division occurs.

step time: A sequencer mode where events are entered one at a time.

stripe: Recording timecode onto a tape.

System Exclusive (Sys-Ex): A type of MIDI data that applies to a specific brand of instrument.

TDM: "Time-Division Multiplexing." TDM allows many different signals to be sent along a common data highway, or bus, at a very fast rate. These signals are "time-sliced" into their own "time-slots," and travel together as a single data stream. When they get to their destination, the receiver can split out the signal(s) it needs.

timbre: Tone color. The quality of a sound that distinguishes it from other sounds with the same pitch and volume.

timecode: A timing reference used to synchronize different audio devices together or to film and video.

tremolo: A cyclic change in amplitude, usually in the range of 7 to 14Hz. Usually achieved by routing a LFO (low frequency oscillator) to a VCA (voltage controlled amplifier).

TRS: Tip-ring-sleeve. TRS phone and mini-phone plugs are used for stereo audio connections, with one channel connected to the tip and the other connected to the "ring" (a metal region between the tip and the sleeve). The sleeve connects to ground. TRS connectors are also used for balanced monaural audio connections; in this setup, the tip and ring both carry the signal. See balanced. They are also used to split out loop send and loop return connections in mixers and signal processors.

undo: Canceling the results of a previous operation, or string of previous operations ("multiple levels" of undo).

VCA: Voltage Controlled Amplifier. A circuit whose gain is determined by a control voltage.

VCF: Voltage Controlled Filter. A filter whose cutoff frequency or resonant frequency is determined by a control voltage.

velocity sensitivity: A keyboard which can respond to the speed at which a key is depressed; this corresponds to the dynamics with which the player plays the keyboard. Velocity is an important function as it helps translate the performer's expression to the music. Velocity data is also transmitted over the MIDI line.

vibrato: A cyclic change in pitch, usually in the range of 7 to 14Hz.

white book: This is sometimes known as karaoke CD, and is used in applications where the combination of full-motion video and audio is needed.

XLR: An industry generic term for round, latching three-pin connectors. XLR connections are used for balanced audio, microphones, and AES/EBU digital signals.

yellow book: The CD-ROM standard for computer data.

Index

Contributors

JIM AIKIN: Senior editor of *Keyboard* magazine.

CRAIG ANDERTON: Coined the term "electronic musician," is a musician and author of various books, including *Home Recording for Musicians, Do-It-Yourself Projects for Guitarists, Electronic Projects for Musicians,* and *MIDI for Musicians*. Craig has also written numerous articles for magazines such as *Guitar Player, Keyboard, EQ, Rolling Stone, Mix,* and *Byte*. In addition to serving as consulting editor to *Guitar Player* and technology editor to *EQ,* Craig lectures all over the world, consults to manufacturers in the music business, and is responsible for some of the sounds you hear coming out of various instruments, as well as some of their design features.

MICHAEL BABCOCK: Writer for *Music & Computers* and other publications.

EDDIE CILETTI: Owns and operates Manhattan Sound Technicians, a service center in New York City. He almost exclusively repairs digital audio tape machines — DAT, ADAT and "DTRS" (Tascam's DA-38, -88 & -98), although he also has a passion for tube gear, juke boxes and roller skating (not blades). You can visit his web site at www.tangible-technology.com or send e-mail to edaudio@interport.com, no question too small.

JULIAN COLBECK: Keyboard player for Steve Hackett, ABWH, Yes, John Miles and Charlie, amongst others. He is the author of more than a dozen books, including the best-selling *Keyfax* series of buyer's guides, and is co-founder of Keyfax Software.

SCOTT GARRIGUS: Multimedia musician, teacher, and writer as well as a certified MIDI maniac.

CHRIS GILL: Music writer who has served as associate editor of *Guitar Player* and editor of *Vintage Gallery*.

JOE GORE: Consulting editor of *Guitar Player* and has recorded and toured as guitarist with Tom Waits, PJ Harvey, and many others.

TED GREENWALD: Editor of *Interactivity* magazine. He has also been an editor for *Guitar Player* and *Musician,* and he is author of *The Musician's Home Recording Handbook*.

DAVID MILES HUBER: Widely acclaimed in the recording industry as a digital audio consultant, author, engineer, university professor, guest lecturer and professional musician. He has authored numerous books on the subjects of recording and electronic music and is a contributing editor for *EQ* magazine.

BRENT HURTIG: Freelance writer specializing in multimedia and music. He is the former editor of *EQ* magazine.

MIKE HURWICZ: Writer and consultant based in Brooklyn, NY.

CAROLYN KEATING: Audio engineer, musician, and freelance editor who has worked for record labels, production companies, and edited numerous music publications.

JEFF KLOPMEYER: Musician as well as editor-in-chief of *First Reflection,* published by Alesis Corp.

ROBERT LAURISTON: Co-author of *The PC Bible* and has written about computer hardware and software for many magazines.

MICHAEL MARANS: Former editor of *Keyboard,* as well as a guitarist and recording authority.

HOWARD MASSEY: Heads up On The Right Wavelength, a MIDI consulting company, as well as Workaday World Productions, a full-service music production studio.

ROGER NICHOLS: Legendary engineer/producer Roger Nichols has been touring with Steely Dan for the last 2 years and has engineered all of their albums. His list of album credits is immense with artists from Gloria Estefan to John Denver, Frank Sinatra, Rickie Lee Jones and Placido Domingo. Roger has three Grammy awards and six Grammy nominations.

DAVE O'NEAL: Plays and records with his band Industrial Soup and writes video game music and software for Electronic Arts.

MARTIN POLON: Internationally syndicated magazine columnist writing about audio, video, multimedia and computing in the pages of *EQ, Studio Sound, One-To-One, Television Broadcast* and numerous other publications. Polon has taught at the University of Massachusetts at Lowell and the University of Colorado as well as being a member of a curriculum body in the recording arts at the University of California.

GREG RULE: Associate editor of *Keyboard.*

MARVIN SANDERS: Editor-in-chief of *Keyboard.*

BENNET SPIELVOGEL: A.k.a. "The East Side Flash," owns Flashpoint Recording Studios in Austin, TX.

MARK VAIL: Associate editor of *Keyboard* and author of *The Hammond Organ.*

GUY WRIGHT: Technical editor of *Interactivity.*

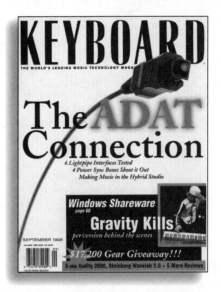